AF871168

# Library Assessment

# SPEC Kits

Supporting Effective Library Management for Over Thirty Years

Committed to assisting research and academic libraries in the continuous improvement of management systems, ARL has worked since 1970 to gather and disseminate the best practices for library needs. As part of its committment, ARL maintains an active publications program best known for its SPEC Kits. Through the Collaborative Research/Writing Program, librarians work with ARL staff to design SPEC surveys and write publications. Originally established as an information source for ARL member libraries, the SPEC series has grown to serve the needs of the library community worldwide.

## What are SPEC Kits?

Published six times per year, SPEC Kits contain the most valuable, up-to-date information on the latest issues of concern to libraries and librarians today. They are the result of a systematic survey of ARL member libraries on a particular topic related to current practice in the field. Each SPEC Kit contains an executive summary of the survey results; survey questions with tallies and selected comments; the best representative documents from survey participants, such as policies, procedures, handbooks, guidelines, Web sites, records, brochures, and statements; and a selected reading list—both print and online sources—containing the most current literature available on the topic for further study.

## Subscribe to SPEC Kits

Subscribers tell us that the information contained in SPEC Kits is valuable to a variety of users, both inside and outside the library. SPEC Kit purchasers use the documentation found in SPEC Kits as a point of departure for research and problem solving because they lend immediate authority to proposals and set standards for designing programs or writing procedure statements. SPEC Kits also function as an important reference tool for library administrators, staff, students, and professionals in allied disciplines who may not have access to this kind of information.

SPEC Kits can be ordered directly from the ARL Publications Distribution Center. To order, call **(301) 362-8196**, fax **(301) 206-9789**, e-mail **pubs@arl.org**, or go to **http://www.arl.org/resources/pubs/**.

Information on SPEC Kits and the SPEC survey program can be found at **http://www.arl.org/resources/pubs/spec/index.shtml**. The executive summary for each kit after December 1993 can be accessed free of charge at **http://www.arl.org/resources/pubs/spec/complete.shtml**.

# SPEC Kit 303

## Library Assessment

December 2007

**Stephanie Wright**

Natural Sciences Information Services Librarian/

Management Information Librarian

University of Washington

**Lynda S. White**

Associate Director, Library Management Information Services

University of Virginia

Association of Research Libraries

Series Editor: Lee Anne George

SPEC Kits are published by the

Association of Research Libraries
21 Dupont Circle, NW, Suite 800
Washington, DC 20036-1118
P (202) 296-2296 F (202) 872-0884
http://www.arl.org/spec/
pubs@arl.org

ISSN 0160 3582

ISBN 1-59407-795-9
978-1-59407-795-1



∞ The paper used in this publication meets the requirements of ANSI/NISO Z39.48-1992 (R1997) Permanence of Paper for Publications and Documents in Libraries and Archives.

# SPEC Kit 303

Library Assessment

December 2007

## SURVEY RESULTS



## REPRESENTATIVE DOCUMENTS

**Assessment Charges and Mission Statements**



**Organization Charts**

## Assessment Web Sites



## Assessment Activity Reports

**Assessment Plans**



## SELECTED RESOURCES

# SURVEY RESULTS

# EXECUTIVE SUMMARY

## Introduction

To assess, in general, is to determine the importance, size, or value of; to evaluate. Library staff assess operations by collecting, interpreting, and using data to make decisions and to improve customer service. They study internal processes, levels and quality of service, and library impact on institutional goals.

The number of assessment activities undertaken in libraries over the last decade has grown exponentially. Libraries of all kinds are looking more closely at how and how well they are serving their users. What may have begun as the occasional assessment duty assigned to the library staff member with the most interest or greatest statistical acumen, has blossomed at many institutions into a formalized library assessment position, committee, department, or all three.

Although this growing area of library management has become recognized as a legitimate use of limited budgets and time, there is not as yet a good overview of precisely how library assessment activities are being implemented or developed. This survey sought to address that missing piece of the puzzle—to examine the current state of library assessment, as well as to provide a starting point for those seeking to develop a library assessment program at their own institutions.

This survey was distributed to the 123 ARL member libraries in May 2007. Seventy-three libraries completed the survey for a response rate of 60%. Only one library indicated that it did not engage in any assessment activities beyond collecting annual data for the ARL statistics, though no reason was given as to why this was the case.

The respondents are primarily from US academic libraries, 63% in public institutions and 22% in private institutions. Twelve percent are libraries in Canadian academic institutions, all of which are public. Public libraries account for only 3% of the respondents. This closely reflects the membership distribution of ARL.

## Assessment Activities

Survey results indicate that while a modest number of libraries in the 1980s and earlier engaged in assessment activities beyond annual ARL statistics gathering, the biggest jump in activity occurred between 1990 and 2004. The overwhelming majority of responses indicate the impetus was service driven and user centered and came from within the library itself rather than from an outside source. Respondents' top impetus for beginning assessment activities (63 respondents or 91%) was the desire to know more about their customers. Based on responses to a question about their first assessment activities, over half began with a survey, almost all of which were user surveys.

It is clear from the survey results that respondents use a wide variety of methods in their assessment endeavors. All respondents have gathered statistics, presumably at least ARL statistics, but every one of the methods listed in the survey has been used by at least one of the respondents at some point, either currently or in the past. The top five assessment methods currently being used are statistics gathering, a suggestion box, Web usability testing, user interface usability, and surveys that were developed outside the library. Locally designed user satisfaction surveys used to be widely used. Now, 20 of the 31 libraries (65%) that previously used this method have switched to surveys developed elsewhere, such as LibQUAL+®. The five least used methods are secret shopper studies, the Balanced Scorecard, wayfinding studies, worklife/organizational climate studies, and unit cost analysis.

The areas of the library being assessed are as varied as the methods used. In the last five years, every function of the library listed in the survey has been assessed by at least one respondent. Almost every respondent has assessed the library's Web site, most frequently with a usability study. Other widely assessed areas include electronic resources, usually assessed by statistics collection and analysis; user instruction, evaluated through statistics and surveys; and reference and collections, both most frequently assessed through statistics collection and analysis. Administrative functions, including human resources, financial services, marketing, and development, that are not centered on users are least evaluated; 30% of the respondents have not assessed even one of these areas.

## Organization of Assessment Activities

Respondents were asked to identify where assessment responsibility fits into their organizational structure. Forty-nine respondents reported that responsibility for assessment activities rests on either a single full- or part-time individual (24 or 34%), an ad-hoc or standing committee (16 or 23%), or a formal department (9 or 13%). All but one of the full- and part-time assessment coordinators and department heads is within two reporting levels of the library director. The remaining 21 respondents (30%) described another organizational structure. The majority of these (15 or 71%) are decentralized, with various units doing their own assessments as needed. For large-scale projects such as LibQUAL+®, an ad-hoc team or committee may be formed. The remaining respondents either use a combination of coordinators and committees or are in the process of creating a new coordinator position.

Though respondents indicated that assessment activities have been performed in their libraries over the last 20+ years, the presence of staff who have primary responsibility for assessment activities has a much more recent history. All but one of the part-time and two of the full-time coordinator positions were created between 2002 and 2007; all of the assessment departments were created in 2000 or later. Nearly 60% of these positions and departments were created between 2005 and 2007. All four ad-hoc committees were created between 2002 and 2007. Standing committees or teams have the longest history of primary assessment responsibility (one since 1984), but the most recent was created in 2007. Departments average 2.4 FTE; committees average six to seven members.

At nearly all of the responding libraries, regardless of organizational structure, assessment staff analyze, interpret, and report on data collected in assessment activities and consult with staff on assessment methods and needs. They frequently perform assessment activities and coordinate the collection and reporting of data. They train staff at just over half of the libraries. They only approve assessment projects at 25% of the responding libraries.

Full- and part-time coordinators and assessment department staff are very similar in the tasks they perform, although part-time coordinators are less likely to be responsible for training staff or monitoring projects. Standing committees are less likely to coordinate the collection, reporting, or archiving of data, to fill requests for library data, or to submit external surveys.

The majority of assessment staff have collaborated on assessment activities with other non-library departments, agencies, or units within the institution, though standing committees are less likely to do so. These non-library collaborations are most often with institutional offices of research and learning, information technology, and assessment and planning.

## Assessment Results Distribution and Outcomes

Methods of distributing assessment results vary depending on the audience, although overall, the most frequently used method is through a Web site. In addition, the methods most widely used to inform the parent institution are print reports and library newsletter articles, while presentations and e-mail announcements are used more frequently for library staff. Staff appears to be the most targeted audience for the distribution of library assessment results; all methods except a campus newsletter are heavily used for them. Results are overwhelmingly distributed to the general public through a Web site or library newsletter articles.

The top two types of assessment information listed on a library's assessment Web site (whether publicly accessible or staff-only) are general library statistics and analyses of assessment activity results. Assessment publications are found more frequently on a public Web site than on a staff-only Web site, while presentations and assessment data are provided more on staff-only Web sites than on public ones. Other types of information mentioned by more than one respondent include meeting notes and agendas on staff-only Web sites.

There is little point in having an assessment program unless the results are used to make improvements in services. Respondents were asked to list three outcomes that were attributable to their assessment activities. Twenty areas were reported, but changes to Web sites and facilities were the most frequently mentioned. Collections, hours, and staff formed the next highest groups. Other areas that were changed include customer service, journals, access services, the online catalog interface, instruction and outreach, and reference services. Only one respondent reported no changes attributable to assessment.

## Professional Development

When asked if their library provides assessment training to library staff, all but 20 of 68 respondents (71%) indicated they received some sort of support for training, whether provided by the library (28%), their institution (32%), or an outside source (62%). When the library provides training, the topics focus primarily on assessment methods, basic statistics, survey construction, the value of assessment, and data analysis.

When evaluating assessment-related professional development venues (such as conferences) outside the institution, the most highly recommended and most attended events were ARL assessment-related meetings and the 2006 Library Assessment Conference. When asked to identify the professional development needs not being met by the aforementioned conferences, respondents focused on training, indicating that there is a lack of available instruction on basic statistical analysis, methodologies, and tools.

## Culture of Assessment

The survey included a series of statements on the culture of assessment. Respondents were asked to rate on a scale of 1 to 5 how well the statements described their respective libraries. Between 68% and 79% of the respondents agreed or strongly agreed with statements related to the commitment of their library administrations to assessment. The remaining statements were related to staff and their support for, or ability to carry out, assessment activities. Only 50% or fewer of the respondents rated these statements at agree or strongly agree; most cluster around the middle of the scale. There appears to be a strong administrative commitment to assessment that does not translate to the organization as a whole.

Just under half of the respondents (31 or 46%) indicated that there is an assessment plan in some or all of their library's departments or units or a library-wide assessment plan. Fifteen respondents commented that the library was either in the process of developing a plan or used an alternate document (such as a strategic plan or annual report) as their assessment plan.

## Conclusions

What do "typical" library assessment programs look like? The typical programs began in the 1990s and engage in various assessment activities in addition to the collection of ARL statistics. They began by doing a user survey because the library wanted greater knowledge of its users and wanted to determine which new services to offer. The programs most frequently gather statistics (100%), but are also strongly involved in doing various user surveys, Web usability testing, and focus groups. They have performed studies of their Web sites. They track usage statistics for electronic resources and assess user education programs, collections, and reference. They have not usually assessed their administrative areas that are not centered on the library user.

Typically, various individual library departments or units do assessment, although the number of institutions with assessment coordinators or committees is growing. The coordinators have typically been appointed within the last five years (2002 to 2007) and are within two reporting levels of the library director. If there is an assessment department, it has just over two members. Committees sometimes date to the 1990s and average six to seven members. The tasks performed by all are remarkably uniform; they analyze, interpret, and report on assessment activities, consult with staff on assessment methods and needs, and perform assessment activities. They coordinate their work with other units in their institutions.

Results of activities are usually distributed through a Web site; they are communicated with staff more frequently than with the parent institution or the general public. Both staff and public Web sites most often present general library statistics and analyses of assessment results. Assessment does lead to programmatic changes in the library, primarily changes to Web sites and facilities.

Training in assessment is supported by the library but is mostly outsourced rather than local. Training that is provided by the library is focused on assessment methods, basic statistics, and surveys. The most highly regarded training appears to come through ARL-sponsored events such as meetings at American Library Association conferences and the Library Assessment Conference. These venues are also appreciated for their networking and sharing opportunities. But more training is needed in assessment basics.

Library administrations are typically committed to the concept of a culture of assessment in their libraries, but there is a perception that this commitment is not shared by all staff. Many staff do not have the skills or rewards needed to carry out assessment projects. Most libraries have an assessment plan or are using a similar alternative document, or they are in the process of developing a plan.

In short, library assessment is alive and well in North America. There has been considerable progress in this area from the mid-1980s through 2007. For that progress to continue, there needs to be more effort to train not only those responsible for assessment, but all staff who are expected to participate in assessment activities.

# SURVEY QUESTIONS and RESPONSES

The SPEC survey on Library Assessment was designed by **Lynda S. White**, Associate Director, Management Information Services, University of Virginia, and **Stephanie Wright**, Natural Sciences Information Services Librarian/ Management Information Librarian, University of Washington. These results are based on data submitted by 74 of the 123 ARL member libraries (60%) by the deadline of June 8, 2007. The survey's introductory text and questions are reproduced below, followed by the response data and selected comments from the respondents.

To assess, in general, is to determine the importance, size, or value of; to evaluate. In libraries, we assess by collecting, interpreting, and using data to make decisions and to improve customer service. We study internal processes, levels and quality of service, and library impact on institutional goals.

The number of assessment activities undertaken in libraries over the last decade has grown exponentially. Libraries of all kinds are looking more closely at how and how well they are serving their users. Since 2004, ARL has sponsored a program to assist libraries with the assessment of services that they offer their users and the processes that support those services. This visiting program has been part of a move away from measuring inputs and outputs and toward judging service quality by measuring outcomes and the value that library programs add to their communities. What may have begun as the occasional assessment duty assigned to the library staff member with the most interest or greatest statistical acumen, has blossomed at many institutions into a formalized library assessment position, committee, department, or all three. These programs may include such activities as statistics collection, conducting surveys, conducting focus groups, Web usability testing, and benchmarking, etc.

Although this growing area of library management has become recognized as a legitimate use of limited budgets and time, there is not as yet a good overview of precisely how library assessment activities are being implemented or developed. The purpose of this survey is to address that missing piece of the puzzle—to examine the current state of library assessment, as well as to provide a starting point for those seeking to develop a library assessment program at their own institutions. This data should help libraries assess where they are and identify best practices to help them build or expand their own programs.

## BACKGROUND

1. Does your library engage in any assessment of library activities (such as statistics collection, conducting surveys, conducting focus groups, Web usability testing, benchmarking, etc.) beyond collecting annual data for the ARL statistics? N=74

| | | |
|---|---|---|
| **Yes** | 73 | 99% |
| **No** | 1 | 1% |

## ASSESSMENT ACTIVITIES

2. Please indicate which of the specific assessment methods below your library is currently using or has used in the past. Check "Currently Used" for methods that the library continues to use to assess activities. Check "Previously Used" for methods that were once used but are no longer used. Check "Never Used" for methods the library has never tried. Check one category for each row. N=73

| Assessment Method | N | Currently Used | Previously Used | Never Used |
|---|---|---|---|---|
| Surveys N=73 | | | | |
| Surveys developed elsewhere (e.g., CSEQ, LibQUAL+®) | 72 | 55 | 15 | 2 |
| Locally designed user satisfaction survey | 70 | 36 | 31 | 3 |
| Facilities use studies | 69 | 41 | 22 | 6 |
| Online user feedback (pop-up windows, etc.) | 69 | 38 | 18 | 13 |
| Worklife/organizational climate studies | 66 | 20 | 21 | 25 |
| Qualitative Methods N=71 | | | | |
| Focus Groups | 70 | 51 | 16 | 3 |
| Secret Shopper Studies | 70 | 5 | 3 | 62 |
| Suggestion Box | 70 | 60 | 6 | 4 |
| Observation | 69 | 37 | 22 | 10 |
| Interviews | 67 | 38 | 22 | 7 |
| Statistics N=73 | | | | |
| Statistics gathering (e.g., e-resource usage, gate counts, ARL statistics, etc.) | 73 | 72 | 1 | — |
| Data mining and analyses | 68 | 43 | 6 | 19 |
| Statistics inventory | 66 | 40 | 9 | 17 |

| Assessment Method | N | Currently Used | Previously Used | Never Used |
|---|---|---|---|---|
| **Usability** N=72 | | | | |
| User interface usability | 72 | 57 | 9 | 6 |
| Web usability testing | 71 | 59 | 9 | 3 |
| Wayfinding | 68 | 15 | 10 | 43 |
| **Other Methods** N=73 | | | | |
| Student learning outcomes evaluations | 70 | 40 | 11 | 19 |
| Benchmarking | 68 | 37 | 11 | 20 |
| Unit cost analyses | 67 | 23 | 25 | 19 |
| Balanced Scorecard | 65 | 4 | 4 | 57 |
| Process improvement | 65 | 29 | 20 | 16 |
| Other method not included above | 31 | 10 | 4 | 17 |

## If you selected "Other method not included above," please specify that assessment method.

### Currently Used

"Administrative Unit Review — review and evaluate library department services or operations."

"Card sorting (usability); heuristic evaluations."

"Card sorts used to gather user input on the library Web site's content structure. Users were provided index cards of keywords that identify library services, collections, facilities, and other related information. They were requested to group the label cards according to about 6 broad categories. A spreadsheet and color coding was used to analysis the frequency of grouped items. This process was used to advise the redesign of a Web site that was user friendly. Clicker Response System. A concept was demonstrated to the class of students. Then a question was shown on their computer screens which presented the same concept in a different situation. They used the Clicker Response System to check how well the students understood the concept and could transfer it to another situation before we moved on to other material."

"Comparison of collection with authoritative lists (a form of benchmarking). Comparison of library collection to what faculty are citing in their publications."

"Faculty Contact Database to track responses to faculty issues and concerns (qualitative and quantitative)."

"In the fall 2006 semester, a class in our Library and Information Science Program, 'Competitive Intelligence and Data Mining (LIS 7490),' used a simulation software package, Arena, to combine data supplied by the University Libraries and observations by students to analyze functions in our reference areas. The simulation and accompanying analysis has resulted in a significant reorganization of the reference area in one of our buildings. The simulation software is now being acquired by the University Libraries."

"Online card sort."

"Outcome measure assessing impact of a grant funded project that digitized special collection material on teaching and learning."

"The library has established a set of metrics corresponding to the goals, objectives, and strategies in the Academic Affairs Strategic Plan for Academic Excellence."

"Usage stats for e-resources."

**Previously Used**

"+/Delta exercise."

"Review of staff organization to look at efficiencies and possible duplication of functions across departments."

"Several years ago we did a telephone survey of student and faculty. It was conducted for us by a survey group within the university. It was moderately useful."

"Use of outside consultants for programmatic reviews."

3. In what year did your library begin assessing library activities beyond the annual ARL data gathering? What was the first assessment activity (survey, focus group, usability test, etc.)? N=63

**Year Assessment Activities Began N=61**

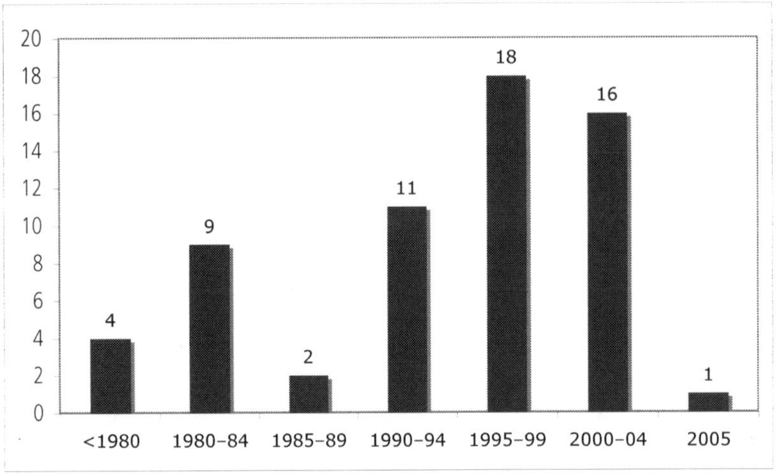


Year Range: 1911 to 2005

**First Assessment Activity N=62**

**<1980**

"ARL's Management Review and Analysis Program."

"Periodic user surveys."

"Survey on material and stack location."

"We have been collecting statistics and doing assessment since the library opened."

**1980–84**

"A satisfaction survey was distributed in paper form to library visitors, mostly students."

"ARL/OMS study."

"At least since 1980, there have been regular calculations of indirect costs of library activities as an associated cost of research. These studies were meant to facilitate university grants requests."

"Cost studies."

"Organizational Assessment."

"Participation in university-wide student satisfaction survey including satisfaction with library."

"Self-study for accreditation. This is the earliest study I can find, but there may be something even earlier."

"Survey of user attitudes: pre- and post opening of a new library facility."

"Work analysis by random sampling."

**1985–89**

"Facility use survey in preparation for construction of addition to main research library building."

"Undergraduate Services Task Force Report entitled 'Preparing Undergraduate Students for the Information Age.'"

**1990–94**

"Exit surveys of library users."

"Locally designed user satisfaction survey."

"Locally devised user survey."

"Organizational climate and organizational structure assessment."

"Survey and focus group."

"Survey of university faculty for: user satisfaction, spending priorities, behavior (frequency of use, etc.)"

"System-wide user surveys."

"User exit survey (behavior, satisfaction, etc.)"

"User satisfaction survey."

"User survey."

"Various customer surveys, comparative peer assessments based on ARL data, various and recurring cost studies."

**1995–99**

"1998-focus groups with staff; 1999-survey of graduate students; 1999-survey of faculty."

"In-house developed user survey."

"Large-scale user survey."

"LibQUAL+™"

"Locally developed comprehensive user survey."

"Not sure, but suspect some form of assessment was undertaken at various times in our history. I am personally aware of having observation, survey, focus group and individual interviewing, and process mapping used as early as 1997."

"SERVQUAL survey of satisfaction of users of undergraduate library."

"Study of print journal 'cost per use.' The library needed to reduce its expenditures on journal subscriptions."

"Survey." (2 responses)

"Survey of users perceptions on service delivery, facilities, service needs."

"There is an indication that a 'Campus Library Assessment' was conducted in 1972, however this cannot be substantiated at this time. In 1995 a comprehensive survey was conducted by the library system."

"User Needs Assessment."

"User satisfaction survey."

"User satisfaction survey in collaboration with regional universities."

"User survey."

"We have always done some assessment: 1 of 6 very least, the 7-yearly program review includes self-study and surveys."

**2000–04**

"Benchmarking for strategic planning."

"Exit survey."

"Focus group."

"Formal Library Assessment Plan."

"LibQUAL+®" (7 responses)l

"LibQUAL+® and Process Improvement Studies."

"Survey."

"Web site design survey."

**2005**

"We have been conducting various assessment activities including unit cost studies, usability studies, focus groups, etc. for many years. We instituted a formal assessment program in 2005 with the first activity being a user satisfaction survey for our Law Library."

## 4. What was the impetus for beginning these assessment activities at your institution? Check all that apply. N=69

| | | |
|---|---|---|
| Desire to know more about your customers | 63 | 91% |
| Investigation of possible new library services or resources | 49 | 71% |
| Desire to know more about your processes | 45 | 65% |
| Desire to identify library performance objectives | 43 | 62% |
| Need to reallocate library resources | 38 | 55% |
| Accountability requirements from your parent institution | 26 | 38% |
| Institutional or programmatic accreditation process | 20 | 29% |
| Proposal from staff member with assessment knowledge | 12 | 17% |
| Other (please specify) | 16 | 23% |

"Comparisons with other institutions, decision-making, resource allocation, strategic planning."

"Desire to facilitate university research activities."

"Desire to focus our attention for making improvements, desire to give authority to our observations when making plans."

"Desire to inform users of services and resources, encourage user input concerning library issues, and document the needs of the library system."

"External review."

"Formative & Summative Evaluation of Educational Resource Development Projects."

"Identifying client focused priorities for service quality and process improvement."

"Initiative from Director of Libraries."

"Need for expenditure reduction."

"Preparation for move to new building."

"Strategic planning."

"To meet a requirement in new Strategic Plan for a user-centered approach to libraries services."

"To meet objectives in the library's strategic plan."

"Too long in the past to be really sure."

"User satisfaction."

"We were planning to open a library expansion and wanted to know more about user satisfaction with services and facilities."

**5. Please indicate which of the following departments/units your library has assessed since 2002 and what methodologies were used for those assessments. Check all that apply. N=67**

| Library Function | N | Surveys | Qualitative Methods | Statistics Collection & Analysis | Usability | Other | Have not assessed |
|---|---|---|---|---|---|---|---|
| Electronic Resources | 66 | 29 | 15 | 56 | 24 | 3 | 2 |
| Interlibrary Loan | 66 | 30 | 10 | 51 | 4 | 4 | 7 |
| Web site | 66 | 32 | 25 | 32 | 59 | 4 | — |
| Branch Libraries | 65 | 40 | 23 | 36 | 7 | 2 | 14 |
| Circulation/Reserve | 65 | 27 | 10 | 49 | — | 5 | 6 |
| Collections | 65 | 25 | 28 | 57 | 9 | 3 | 3 |
| Reference | 65 | 37 | 26 | 53 | 3 | 2 | 5 |
| User Instruction | 64 | 45 | 24 | 46 | 3 | 4 | 3 |
| Digital Initiatives | 63 | 24 | 17 | 33 | 21 | 1 | 16 |
| Online Catalog | 63 | 25 | 11 | 33 | 34 | 2 | 6 |
| Acquisitions | 62 | 9 | 8 | 46 | — | 8 | 13 |
| Cataloging | 62 | 3 | 6 | 43 | 1 | 9 | 15 |
| IT Systems | 62 | 12 | 13 | 27 | 11 | 3 | 20 |
| Shelving | 62 | 11 | 6 | 45 | 1 | 4 | 12 |
| Facilities | 61 | 35 | 20 | 27 | 6 | 6 | 6 |
| Human Resources | 61 | 8 | 12 | 10 | — | 1 | 41 |
| Preservation | 61 | 8 | 8 | 35 | 1 | 5 | 20 |
| Publicity/Marketing | 61 | 10 | 9 | 4 | 1 | 1 | 42 |
| Selectors/Subject Liaisons | 61 | 22 | 19 | 29 | — | 2 | 16 |
| Special Collections | 61 | 14 | 14 | 32 | 3 | 4 | 19 |
| Administration | 60 | 9 | 15 | 12 | — | 4 | 31 |
| Financial/Business Services | 60 | 5 | 4 | 21 | 1 | 5 | 32 |

| Library Function | N | Surveys | Qualitative Methods | Statistics Collection & Analysis | Usability | Other | Have not assessed |
|---|---|---|---|---|---|---|---|
| Staff Training/ Development | 60 | 37 | 14 | 14 | 1 | 1 | 17 |
| Work Climate | 60 | 29 | 18 | 4 | 1 | — | 23 |
| Development/ Fundraising | 59 | 3 | 5 | 22 | — | — | 36 |
| Other | 16 | 3 | 2 | 2 | 1 | 4 | 9 |

Please specify other library function that was assessed.

| Library Function | Assessment Method |
|---|---|
| Acquisitions & Cataloging | Process improvement |
| Book availability | (not specified) |
| In general, all of our units are assessed through our user surveys. | |
| Information Literacy | (not specified) |
| Libraries; Technical Services | In Summer 2006, an external review committee was appointed to assess the libraries (as a unit). In Summer 2007, a consulting firm was hired to assess Technical Services |
| Library Spaces | Observational studies and surveys are used to inform decision about construction and renovation. ('Facilities' means maintenance and cleaning activities to me.) |
| Media Services, Special Collections | We conducted a self-study for Special Collections and hired an outside management consultant to assess. We are currently conducting a self-study for Media Services and will have a consultant come in this summer. Also we participated in LibQUAL+™ in 2001 and 2005. |
| Overall with LibQUAL+™ | |
| Public Programs; Government Documents | Public Programs: surveys. Government Docs: statistics and randomized and selective shelf-reading for quality control and improvement. |
| Student Multimedia Design Center planning | The use of multimedia by faculty and students was assessed as part of the planning process for design and implementation of the new 15,000 square foot Student Multimedia Design Center with 70 workstations, 6 studios and 2 classrooms. The assessment included focus groups, surveys and interviews. |
| User satisfaction with online finding tools | |
| Workflow in Cataloging and Acquisitions; Inventory of facilities; audit processes of business office. | |

**Please specify other assessment method(s) that was used.**

"Administrative Unit Review."

"ARL ILL cost/performance for all years."

"Assorted consultant review and reports based on data, interviews, work flow analysis, review etc."

"Branch Libraries: Visits to other libraries to develop benchmarks. Web site: Card sorting tests on technology and organizations. Staff Training Development: Informal conversations with individual staff. Financial/Business Services: Various operational aspects of library business office are periodically assessed by university business services unit."

"Card Sorts, SWOT analysis."

"Collections: WorldCat Collection Analysis. IT Systems: informal feedback and in-house assessments. Cataloging: process improvement. ILL: unit costs."

"For Preservation and Special Collections: assessment of the collections. For many services, we use informal feedback and our suggestion box."

"Gap surveys."

"Informal evaluation of workflow and procedures."

"In-house reviews: data/cost data."

"Observational study of facilities and laptop use; external consultants."

"Process improvement studies."

"Process improvement/CQI (for all six departments/units for which OTHER was checked)."

"The 'qualitative methods' used includes significant individual interactions with students and faculty to ascertain the value of the service and any suggested improvements to services, systems and collections."

"We are at an early stage of using 'student learning outcomes evaluations' to measure the impact of our user instruction. We made use of the simulation software package, Arena, to analyze reference services in several of our buildings."

"We have also done benchmarking studies of our shelving process (1999) and of our training program (2000)."

## ORGANIZATION OF ASSESSMENT ACTIVITIES

6. Who has primary responsibility for coordinating and/or planning your library's assessment activities? Check the one item below that best describes your organization. N=70

| | | |
|---|---|---|
| A single individual working part-time as an assessment coordinator | 13 | 19% |
| A standing committee(s)/team(s) that is charged with assessment | 12 | 17% |
| A single individual working full-time as an assessment coordinator | 11 | 16% |
| A department/unit that is charged with assessment | 9 | 13% |
| An ad hoc committee that is charged with assessment | 4 | 6% |
| Other | 21 | 30% |

7. For which functions below is the assessment coordinator responsible? Check all that apply. N=49

| | Part-Time Individual N=13 | | Standing Committee N=12 | | Full-Time Individual N=11 | | Assessment Department N=9 | | Ad hoc Committee N=4 | | Total N=49 | |
|---|---|---|---|---|---|---|---|---|---|---|---|---|
| | # | % | # | % | # | % | # | % | # | % | # | % |
| Analyzes, interprets, and reports on data collected in assessment activities | 12 | 92% | 11 | 92% | 11 | 100% | 9 | 100% | 4 | 100% | 47 | 96% |
| Consults with staff on assessment methods and needs | 12 | 92% | 11 | 92% | 11 | 100% | 9 | 100% | 3 | 75% | 46 | 94% |
| Performs assessment activities | 11 | 85% | 9 | 75% | 11 | 100% | 9 | 100% | 3 | 75% | 43 | 88% |
| Monitors/coordinates assessment projects throughout the library | 7 | 54% | 11 | 92% | 10 | 91% | 8 | 89% | 2 | 50% | 38 | 78% |
| Coordinates collection of data across the library | 11 | 85% | 5 | 42% | 11 | 100% | 7 | 78% | 3 | 75% | 37 | 76% |
| Submits external surveys (ARL, ALS, NATC, American Library Directory, etc.) | 10 | 77% | 5 | 42% | 10 | 91% | 8 | 89% | 1 | 25% | 34 | 69% |
| Coordinates the reporting/archiving of the library's statistical data | 12 | 92% | 2 | 17% | 7 | 64% | 9 | 100% | 3 | 75% | 33 | 67% |
| Fills requests for library data | 11 | 85% | 2 | 17% | 9 | 82% | 9 | 100% | 2 | 50% | 33 | 67% |
| Provides training on assessment topics | 5 | 38% | 7 | 58% | 8 | 73% | 6 | 67% | 1 | 25% | 27 | 55% |
| Approves assessment projects throughout the library | 3 | 23% | 1 | 8% | 4 | 36% | 2 | 22% | 2 | 50% | 12 | 25% |
| Other, please specify | 4 | 31% | 1 | 8% | 3 | 27% | 5 | 56% | — | — | 13 | 27% |

Please specify other responsibility.

**Part-time Assessment Coordinator**

"Internal staff training. Coordinate searches for professional librarians."

"Leads the library's Assessment Team."

"Serves on Institutional Research Committees."

"Training Coordinator handles most training, and is a member of the Assessment Working Group. I provide guidance and best-practices."

**Standing Assessment Committee/Team**

"Feedback from contributors (assessing the assessment process)."

**Full-time Assessment Coordinator**

"Ethics review applications."

"Supervise student assessment and usability assistants."

"The position is not officially designated as a 'coordinator' although it is an inherit aspect of the position. It is still under development based upon determining the extent and type of assessment coordination needs for the campus library system."

**Assessment Department/Unit**

"Chairs Library Assessment Group; Ex-officio member of the Strategic Planning team and develops strategic planning performance measures and benchmarks."

"Collaborates with the Public Services Executive Committee's Usability & User Studies Committee. The U&US committee was created in 2005, and currently has 8 members. To date, this group's focus has been to promote and facilitate usability within the library. This group: Consults with staff on usability methods and needs. Facilitates usability projects throughout the library. Performs priority usability activities. Analyzes, interprets, and reports on data collected in usability activities. Provides training through priority usability projects. Has provided other training opportunities by reporting on its work and by inviting guest speakers. U&US has been collaborating with the IRB, and may work with the usability group in Cornell's IT department. Other general information: CUL conducts LibQUAL+™ periodically, centrally. Each subject library determines needs and assessment approaches independently. Project-based assessments are conducted on-demand by RAU based on priority. Annual statistics collected/compiled centrally from units."

"Develops programmatic, strategic, and sustainable approach to library assessment activities."

"Manages institutional data repository and development of same. Collaborates with library central IT unit on repository architecture and data structures. Manages development of report writing applications as part of MIS program. Point of contact for IRB. Liaises with university's department of planning and analysis."

"The 'Evaluation and Analysis' unit is part of the Organizational Services department. Organizational Services provides 'back office' services to both the library and the 'Computing and Communications Services' unit. Thus, 'Evaluation & Analysis' also serves the computing unit on campus."

8. Does this position or group collaborate on assessment activities with other non-library departments, agencies, or units within the institution? N=49

| | | Yes | |
|---|---|---|---|
| | N | # | % |
| Part-time Assessment Coordinator | 13 | 9 | 69% |
| Standing Assessment Committee/Team | 12 | 5 | 42% |
| Full-time Assessment Coordinator | 11 | 8 | 73% |
| Assessment Department/Unit | 9 | 8 | 89% |
| Ad hoc Assessment Committee/Team | 4 | 3 | 75% |

If yes, please list the department(s), agency, or unit(s). N=49

**Part-time Assessment Coordinator**

"Budget Office, Government Affairs, Foundation and Government Grants."

"Have had the opportunity to work with one academic department on a survey this year. (School of International and Public Affairs.)"

"Human Resources, Office of Affirmative Action and Diversity Programs, Council on the Status of Women."

"Institutional Planning and Budgeting — share relevant assessment analyses; Libraries provide library-related information to IPB for university-wide surveys; Libraries coordinate running of LibQUAL+™ not to conflict with a graduate student survey run by IPB in same term."

"Provost's office often requests data for higher ed surveys."

"Research, Planning and Institutional Effectiveness."

"University administration — provides statistical reports and participates in accreditation data-gathering."

"University Administration, especially for accreditation; academic departments, especially those undergoing certification and accreditation."

"University Planning and Analysis."

**Standing Assessment Committee/Team**

"Academic Affairs, Advancement, Institutional Research, Washington Research Library Consortium."

"Office of Information Technology. Educational Testing Center, Statistical Consulting Service, Registrar's Office."

"Office of Institutional Research."

"Primarily data exchange with the Office of Institutional Research."

"Statistical analysis."

**Full-time Assessment Coordinator**

"At the university, Libraries and Informational Technology are integrated. This position reports directly to the Vice Provost for Information Services and is also responsible for coordinating the same assessment activities in IT. The position collaborates with the university's Office of Institutional Research and Planning."

"Auditing, Student Life, University Communications, Southern University Libraries."

"Institutional Planning and Assessment, Center for Teaching and Learning."

"Office of Institutional Research."

"Office of Institutional Research & Assessment."

"The Assessment Officer is a member of the University Assessment Committee."

"University office of Institutional Research, Office of Assessment."

**Assessment Department/Unit**

"Center for Teaching Advancement and Assessment Research, Assessment Council Office of Institutional Research and Academic Planning, Various assessment committees working on self-study for 2008 re-accreditation."

"Computing and Communications Services as well as the university's 'analysis and planning' unit."

"Office of Educational Assessment; Office of Learning Technologies."

"Office of Institutional Planning and Analysis, individual faculty."

"RAU will also collaborate with our office of institutional research, IRB, Survey Research Institute and other appropriate units."

"University Assessment and Testing."

"University Institutional Assessment and Studies."

**Ad hoc Assessment Committee/Team**

"Office of Institutional Research."

"Planning & Institutional Research."

"University Committee on Assessment & Institutional Improvement. The Library Director chairs the Sub-Committee on 'University Community Experience.'"

## Part-time Assessment Coordinator

9. Please provide the following information about the part-time assessment coordinator: position title, year position took on assessment responsibility, by how many reporting levels the part-time assessment coordinator is removed from the library director (e.g., Director > Dept Head > Assessment Coordinator = 2). N=13

| Position Title | Year | Reporting Levels |
|---|---|---|
| Director for Planning and Research | 1995 | 2 |
| Training, Assessment & Statistics Coordinator | 2002 | 2 |
| Associate University Librarian | 2003 | 1 |
| Special Assistant to the Director for Programs | 2003 | 1 |
| Coordinator of Assessment & Staff Development | 2004 | 1 |
| Director, Access Services | 2004 | 2 |
| Assessment Librarian | 2005 | 2 |
| Deputy University Librarian | 2005 | 1 |
| Assistant Dean | 2006 | 1 |
| Associate Dean for Organizational Development | 2006 | 1 |
| Coordinator of Assessment | 2006 | 2 |
| Program Coordinator | 2006 | 1 |
| Program Coordinator for Marketing and Assessment | 2006 | 3 |

### Year Position Took on Assessment Responsibility

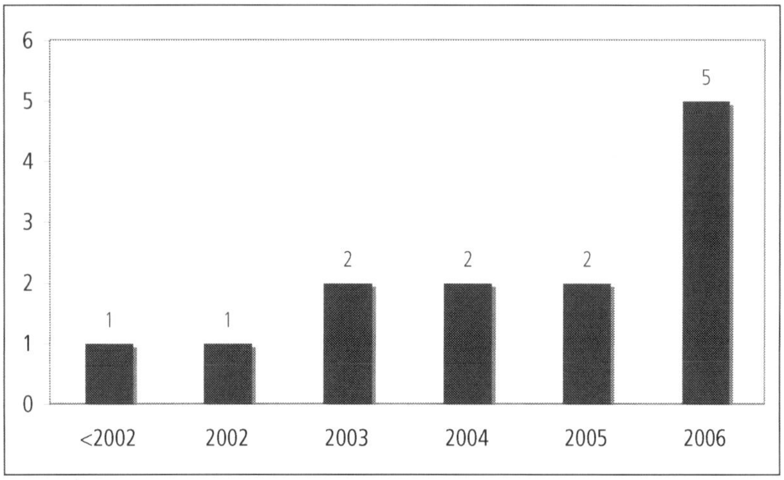


### Reporting Levels

| | | |
|---|---|---|
| One level | 7 | 54% |
| Two levels | 5 | 38% |
| Three levels | 1 | 8% |

## Standing Assessment Committee/Team

10. Please provide the following information about the standing assessment committee/team: name of standing committee/team, position title of standing committee/team leader, year standing committee/team was created, number of standing committee/team members. N=12

| Name of Standing Committee/Team | Position Title of Leader | Created | Members |
|---|---|---|---|
| The Administrative Group plus the Manager of Staff Technology Training & Development and Strategic Planning | University Librarian | 1984 | 6 to 8 |
| Community Needs Assessment Committee | Current team leader is Head, Information Services & Resources Department | 1995 | 7 |
| User Feedback Committee | Co-chaired by reference librarian and collection development librarian | 1995 | 8 |
| USER Team | Director, Collections Services | 1997 | 5 |
| Assessment Working Group | Varies; chair is elected | 1998 | 9 |
| Assessment Committee | Assistant to the Director | 2001 | 12 |
| Assessment Committee | History and Area Studies Librarian | 2002 | 8 |
| Library Assessment Task Force | Manager, Circulation Services | 2003 | 5 |
| Library Services Assessment Committee | Associate Dean | 2005 | 6 |
| University Libraries Assessment Team | Director, Library Resource Management | 2005 | 3 |
| Assessment Committee | Department Chair | 2006 | 6 |
| Library Assessment Working Group | Chair | 2007 | 6 |

### Year Standing Committee/Team Was Created

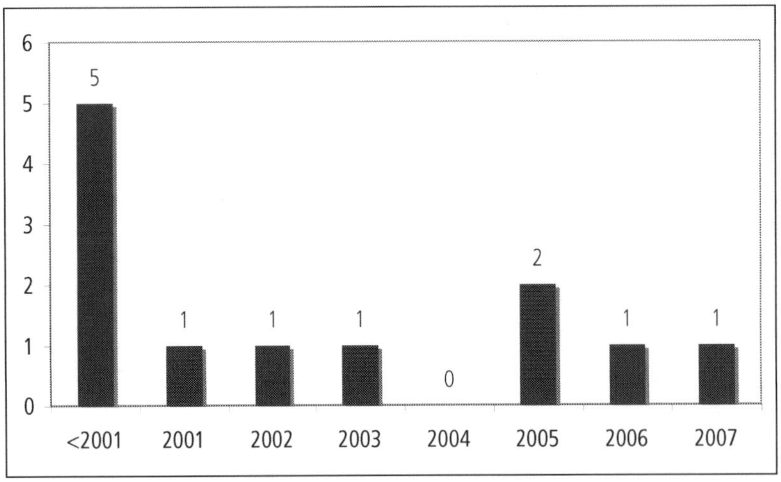

## Full-time Assessment Coordinator

11. Please provide the following information about the full-time assessment coordinator. Position title, year position was created, by how many reporting levels is the full-time assessment coordinator removed from the library director? (e.g., Director > Dept Head > Assessment Coordinator = 2) N=11

| Position Title | Created | Reporting Levels |
|---|---|---|
| Assessment Coordinator | 1999 | 1 |
| Process Improvement Specialist | 2000 | 2 |
| Staff Development Librarian | 2002 | 2 |
| Director of Assessment and Outreach | 2003 | 1 |
| Assessment Librarian | 2004 | 1 |
| Library Assessment Coordinator | 2005 | 2 |
| Director of Planning, Assessment, and Research | 2006 | 1 |
| Associate Dean for Assessment, Personnel and Research | 2006 | 1 |
| Senior Program Office for Research & Analysis | 2006 | 2 |
| Assessment Officer | 2006 | 1 |
| Assessment & Public Services Librarian | March 2007 | 2 |

### Year Assessment Coordinator Position Was Created

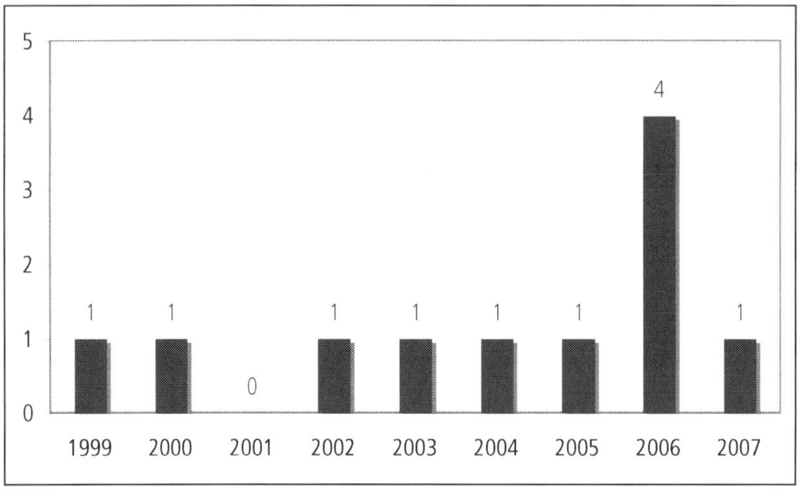


### Reporting Levels

| | | |
|---|---|---|
| One level | 6 | 55% |
| Two levels | 5 | 45% |

## Assessment Department/Unit

12. Please provide the following information about the assessment department/unit. Name of department/unit, position title of department head, year department/unit was created, number of staff in the department/unit, by how many reporting levels is the department/unit head removed from the library director? (e.g., Director> Dept Head>Assessment Department/Unit Head = 2). N=9

| Name of Department/Unit | Department Head | Staff | Created | Levels |
|---|---|---|---|---|
| Evaluation & Analysis | Manager | 1.5 FTE | 2000 | 2 |
| Direction générale – communications | Adjointe au directeur général | 4 | 2000 | 1 |
| Management Information Services | Director | 3 | 2000 | 2 |
| Dean's Office | Associate Dean for Planning & Assessment | 2 | 2005 | 1 |
| Research & Assessment Unit | Director of Service Innovations & Resource Planning | 3.2 devoted to assessment | 2006 | 2 |
| Management Information Services and Communication | Assistant Director | 2.75 FTE | 2006 | 1 |
| Planning and Organizational Research | Associate University Librarian | 2 | 2006 | 1 |
| Office of Assessment & Planning | Director of Assessment & Planning | 1.5 FTE | 2006 | 2 |
| Assessment | Head of Access Services and Assessment | 2 | 2007 | 2 |

### Year Assessment Department/Unit Was Created

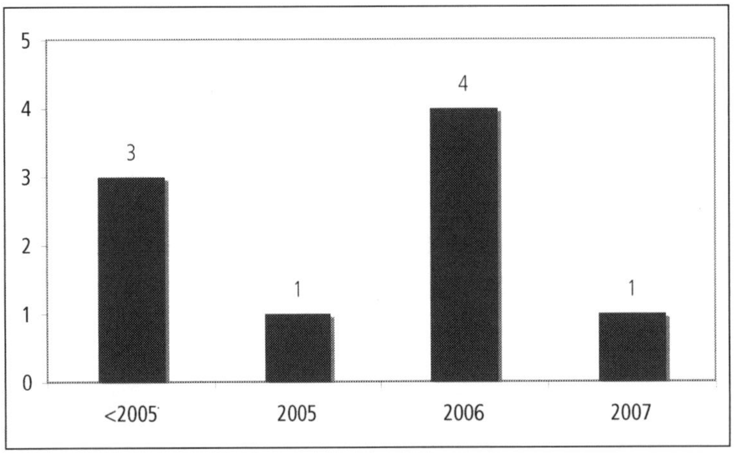


### Reporting Levels

| | | |
|---|---|---|
| One level | 4 | 44% |
| Two levels | 5 | 56% |

## Ad hoc Assessment Committee/Team

13. Please provide the following information about the ad hoc assessment committee/team: Name of ad hoc committee/team, position title of ad hoc committee/team leader, year ad hoc committee/team was created, number of ad hoc committee/team members. N=4

| Name of Committee/Team | Committee/Team Leader | Created | Members |
|---|---|---|---|
| Assessment Team | Director | 2002 | 3 |
| Assessment Committee | Engineering Reference Librarians | 2004 | 6 |
| LibQUAL+™ User Assessment Group | Department Leader, Administrative Services | 2006 | 6 |
| Assessment Committee | Director, Access, Information & Research Services | 2007 | 8 |

## Other Assessment Activities Organization

14. Please briefly describe the organization of assessment activities in your library. N=21

"All ad hoc and as needed."

"Assessment activities have been initiated, conducted, and coordinated at the unit or department level, or by a committee charged for a specific project."

"Assessment efforts have been coordinated through a combination of participants: Library Administration, an Assessment Committee, and a part-time Statistical Data and Assessment Specialist."

"Assessment in the University Libraries is currently distributed among a number of individuals, departments, committees, and task forces. Assessment activities are currently being examined and/or conducted by the: Statistics, Metrics, and Survey Development Committee and subcommittees; the Public Services Committees and associated task forces; and the Collections Committee and associated task forces. The groups communicate with each other regularly."

"Assessment is generally done at the unit/function level. For large-scale assessment efforts such as LibQUAL+™, an ad hoc committee has been appointed."

"Assessment is generally undertaken by a unit within the library, such as the Web Services Group, and has a particular focus. At times, an ad hoc committee is formed for an overarching assessment project."

"Assessment occurs largely in a decentralized way. Individual libraries or organizational units initiate activities. Several system-wide activities have been done—two rounds of LibQUAL+™, a time/cost study, Web usability. These have been overseen by individual ad hoc committees created for that purpose."

"Associate Dean for Research & Access oversees the Biennial LibQUAL+™ Survey and any other surveys conducted. A Usability Studies Task Force, reporting to the e-Library Oversight Committee, plans and implements usability testing for the library Web site, online catalog, etc."

"Combination of: newly formed User Feedback Committee (est. 2007); individual with responsibility for managing and reporting ARL, institutional, and other library-wide data; and ad hoc activity by individuals, groups, and units, as appropriate."

"Information Access and Delivery Service Department is responsible for collection of data within the department the organization and presentation of the data."

"Leadership of library-wide assessment activities is at the administrative level Director/Assistant Director. Leadership of library instruction related assessment is at the department level. Leadership of Web site assessment is as the Assistant Director level. Leadership related to multimedia needs was assessed by the Assistant Director for Library Public Services."

"Oversight for assessment is the responsibility of the associate dean for planning and administration. Under consideration is a half-time position of evaluation and assessment librarian, reporting to this associate dean."

"Participated in LibQUAL+™ Survey January 2007 for the first time. Other: ad hoc, periodic reviews for specific purposes. Assessment Librarian position is currently at job description stage. ARL and other statistics: coordinated by Statistics Coordinator (now Assessment Librarian). Teams (subcommittees) are being formed to assist Assessment Librarian with various tasks."

"Program managers (aka department heads) are expected to have assessment skills & experience to evaluate the success of their services, operations and projects."

"Statistical compilations coordinated by Administrative Office. Focus groups, usability studies surveys designed and implemented by the specific public service group responsible."

"The Assistant to the Dean of Libraries handles assessment, particularly LibQUAL+™ on a part-time basis. A reference librarian does a considerable amount of LibQUAL+™ data analysis and benchmarking. The Associate Director and other library staff have done numerous focus groups over the years, particularly as they relate to facilities and library as place."

"There is a Usability Librarian, others in collections and public services as well as department libraries who conduct assessments, also consultants brought in to conduct university and library surveys [especially in HR activities]."

"Two part time assessment librarians reporting to the director. (We also have an assessment committee.)"

"Various committees (and ad hoc task groups) and various library departments determine and carry out own assessment activities. Some assessment activities are library wide and coordinated by admin of library."

"Very minor other than LibQUAL+™."

"We are just establishing central roles for assessment. Individual departments have done various assessments. Statistics reporting is being centralized in the Administrative Office."

## ASSESSMENT RESULTS DISTRIBUTION

15. What methods are used to distribute data/analysis/results of library assessment activities? Check all that apply. N=69

| | N | To Library Staff | To Parent Institution | To General Public | Not Used |
|---|---|---|---|---|---|
| Presentations | 66 | 58 | 32 | 11 | 2 |
| Print reports (e.g., annual report) | 65 | 49 | 39 | 18 | 8 |
| Web site | 65 | 56 | 40 | 39 | 5 |
| E-mail announcements | 64 | 58 | 16 | 4 | 6 |
| Library newsletter articles | 62 | 45 | 35 | 27 | 11 |
| Campus newsletter articles | 59 | 16 | 31 | 14 | 24 |
| Other | 12 | 4 | — | — | 8 |

Please specify other distribution method(s).

"Attend departmental meetings to report on assessment results."

"In the recent past formal reports on assessment have not been developed exclusively for the 'parent organization,' staff, users or public, but rather recently have been included in internal administrative reports. LibQUAL+® results have been provided to staff via presentations and brief articles describing outcomes have been reported in campus newspaper/e-news outlets. Currently an assessment and evaluation Web site is under development for internal resource purposes."

"Our basic statistical data is accessible through our library Web site."

"Presentations at professional library meetings."

"Staff intranet site."

"The Libraries use a central announcements blog which is available as an RSS feed to any subscribers. Library liaisons to academic departments also share selected assessment information with their assigned departments."

"To the profession through presentations and publications."

"WIKI. Annual budget submission includes annual assessment highlights."

16. If your library has either a staff-only or publicly accessible library assessment Web site, please indicate which kind of information is published there. Check all that apply. N=58

| | N | Staff-only Web site | Publicly accessible Web site |
|---|---|---|---|
| General library statistics | 54 | 40 | 31 |
| Analysis of assessment activity results | 49 | 36 | 27 |
| Presentations | 40 | 30 | 18 |
| Assessment data | 39 | 32 | 16 |
| Publications | 34 | 20 | 25 |
| Online assessment tools (e.g., surveys) | 31 | 26 | 12 |
| Links to other library assessment sites or information | 30 | 22 | 13 |
| Other | 5 | 4 | 2 |

Please specify other information that is published on the staff-only Web site.

"A bibliography of assessment resources, and a white paper on the library as place."

"Assessment group meeting notes and agendas; Working documents."

"Individual comments from USER surveys are on staff-only Web site. Comments gathered from our online comments link as well as comments gathered from our comments boxes are on staff-only Web site."

"LibQUAL+™ results."

"Links to internal department stats and assessment committee membership and contact info."

"List of user focused assessment projects in progress and completed; list of staff focused assessment projects in progress and completed; Assessment Committee Plan (aligned with the libraries' strategic plan); notes of Assessment Committee meetings."

"Minutes of Assessment team meetings, form and procedures for staff to submit ideas for assessment activities."

"Minutes of meetings, annual reports, assessment inventory."

"Reports conducted by external evaluators; guidelines for statistics collection."

"The library assessment plan."

"The staff-only Web site is currently under construction but the intended content, initially, will centrally communicate: the status of assessment initiatives, calendar planned projects, report outcomes, identify working groups/committees, identify administrative priorities, and include reading resources organized by functional areas."

Please specify other information that is published on the publicly accessible Web site.

"Annual Report (ARL rankings); Strategic Planning Metrics."

"Assessment committee membership and contact info."

"LibQUAL+™ results. Lists of specific actions taken in response to user suggestions and assessment data."

"Meeting Minutes, Committee Annual Reports."

"Terms of reference, membership, minutes, suggested readings, project form."

"The library assessment plan."

## ASSESSMENT OUTCOMES

17. Please describe up to three demonstrable outcomes that have been made to your library's programs, policies, or services based on information collected via assessment activities. N=64

Respondent 1

"2006 LibQUAL+™ data motivated OSUL to become a development partner with OCLC for WorldCatLocal as a potential replacement for local OPAC."

"2006 LibQUAL+™ data motivated the change from title to keyword as default OPAC search."

Respondent 2

"A library expansion project was moved up on the campus master plan due to recent facilities' studies which have shown that our main library and our remote storage facility are at 95% capacity and that we will be completely out of shelf space for library collections in 2009."

"Improved delivery of Instant Messaging service to library patrons based on surveys of staff providing the service."

"Increased hours of operation during winter intersession with minor budget implications. The need was conveyed from students in focus groups."

Respondent 3

"A strong user-centered focus in all our initiatives and services."

"Changing of hours in response to LibQUAL+™ data; had immediate results in terms of satisfaction of users."

"Aggressive e-resources collection development."

Respondent 4

"Acquired and made available federated searching for subscribed databases."

"Cancelled low and retained high use databases and journal packages."

"Revamped brochures and established a marketing team."

Respondent 5

"Additional custodial staff assigned to Western Libraries."

"Major renovations funded by the University to increase user-designated space to accommodate increased undergraduate enrollment."

"Purchases made in response to requests."

Respondent 6

"Addressed problems with ILL fill rates and turn around times."

"Improved design and ease of use of library Web site."

"Improved design and ease of use of library online catalog."

Respondent 7

"Adopted measurable methods to assess student learning."

"Library became involved in planning and implementing the Freshmen Experience with extensive information component."

"Quality of service survey informed management of service points in need of improvement."

Respondent 8

"Advocacy for renovating outdated library spaces using the results of the LibQUAL+™ survey. Library users gave it a very high priority, and made many negative comments about existing facilities."

"Journal cancellations and additions resulting from a faculty survey."

'Redesigned library Web site and new library catalog incorporated usability testing results and user feedback. Also included a change to a new chat reference product."

Respondent 9

"Assessment of faculty and student needs related to use of the multimedia resources and facilities included surveys, focus groups and interviews, and resulted in the planning and implementation of a very successful 15,000 square foot Student Multimedia Design Center in 2007 which has been widely praised by faculty and students for its design and functionality and resulting in a heavily used facility."

"Assessment of the library Web site is currently underway and has involved surveys, focus groups, web usability studies and interviews, and the results so far have resulted in many new ideas for the revised Web site to be developed in 2007/2008."

Respondent 10

"Attempts to 'calm' the physical environment of the library in terms of noise/food/inappropriate behavior."

"Experimentation, as funding became available, with 24/5 hours for the main research library."

Respondent 11

"Based on LibQUAL+™ results we decided to stay open 24/7 during dead week and finals week."

"Based on benchmarking selected ARL statistics data against comparator institutions we received a budget increase from the University Administration."

"Our Organizational Climate survey produced a staff development needs assessment; weekly reports from the library administration published in the staff bulletin; department heads attending a 'Strategies for Change' workshop; and a re-structuring of the student employment budget."

Respondent 12

"Based on user satisfaction survey responses a program for customer service training was implemented for training library staff and student employees."

"Based on responses to an organizational culture survey, the internal staff Web site is being moved to a content management site and is going through a complete overhaul with new rules and guidelines for department/unit sites."

"For years our user community has been asking for the ability to use institutional ID cards that link to money accounts to pay for copies and printing. We worked with the copy & print vendors to come up with software to make it possible and will be implementing this change this summer."

Respondent 13

"Better Web site design."

"Identification of strategic initiatives within current strategic plan."

"Improvement to photocopy services."

Respondent 14

"Changed approval plan vendor based on evaluation of profile, return rates, and focus groups with selectors. Improved efficiency of approval plan process to allow for purchase of shelf-ready materials."

"Assessment of budgeting process has led to a built-in replacement cycle for the library's computer equipment."

"Evaluation of software used for instant messaging in virtual reference led to a change from one system to another, more popular and commonly used program. This has led to a real increase in the usage of our chat reference service."

Respondent 15

"Changed library hours based on gate counts and student feedback."

"Increasing expenditures for electronic resources based on LibQUAL+™ findings."

"Redesigning the library Web site based on LibQUAL+™ comments."

Respondent 16

"Coffee shop created in 24-hour study lounge."

"Increases in fundraising and number of donors; and heightened campus awareness of library."

"Increased attention to service quality."

Respondent 17

"Collapsed six service points on entry-level floor of main library to two based on input gathered from assessment activities, asking to simplify user experience."

"Added cafe to entry level of main library based on user input gathered from assessment activities."

"Placed group study rooms on online reservation system based on user input from assessment activities."

Respondent 18

"Currently redesigning public Web pages based on user survey feedback."

"Currently investigating improvements to OPAC based on user survey feedback."

"Currently providing free document delivery based on a combination of user feedback, and results from a survey of peer institutions."

Respondent 19

"Customer Service Academies were held for both professional and paraprofessional staff. A regular program of customer service training was also established for the student assistants working in the library."

"Our Interlibrary Loan Department was reorganized to improve customer service. A program to Buy-not-Borrow certain returnable (i.e., book) requests was established to provide such materials more rapidly."

"The University Libraries' Web site was completely redesigned to allow greater ease of use and accessibility."

Respondent 20

"Development and implementation of service training workshops for library staff and student employees."

"Improvements to library Web site, digital publishing, and presentation of electronic resources."

"Development of plans for renovation to and services in Undergraduate Library."

Respondent 21

"Development of the Assessment Librarian position for the purpose of developing and advancing assessment initiatives that inform library administrative decisions on collection resources, public service and related issues."

"Development of the 'Reference Management Taskforce' (RMT) as a result of the Reference Service Delivery User Survey in 2005. The RMT works to coordinate service related initiatives, facilitate dialogue among campus reference librarians, and identify public service goals and priorities that inform library administration. The survey facilitated the selection of a new virtual IM/Chat reference software and efforts to revise the current reference service staffing model."

"One of the campus libraries used a product Clicker Response System from E-Instruction to collect user data in library instruction sessions. The tool was used in about 40 instruction sessions over the course of a semester. The system of the clicker software was used to receive immediate feedback in class as to the percentage of students that answered questions correctly, incorrectly, or were unsure. The feedback within the class was used to inform the content of the class (go over the concept again or move on to new concepts). The accumulated data from all sessions after the semester was used to identify and prioritize difficult concepts for students and work with faculty and students to find teaching and learning solutions."

Respondent 22

"E-journal subscription decisions based on use and cost analysis. Cancellation of print materials after use study."

"Learning Commons development largely informed by multiple assessment activities. Choices about furniture, technology, and hours were all shaped by iterative assessment efforts."

"Stack and Circulation statistics shaped decisions about off-site storage of materials."

Respondent 23

"Enhancements to role of library liaisons. This came about when focus group data indicated faculty was looking for improved communication and a closer working relationship."

"Identification of quiet study areas within the library as a result of comments from LibQUAL+™ survey."

"Changes in assignment of study carrels based on LibQUAL+™ and focus group data indicating a need for more individual study areas."

Respondent 24

"Established an Information Commons for collaborative learning, which success has resulted in expanding the commons to satellite locations throughout the library."

"From the results of a collaborative study on library usage, casual study areas were enhanced, the number of public workstations were doubled, and application software was installed to match the suite used in general-purpose university computer labs."

"Web usability studies have resulted in a simplified Web site design to reflect patron usage patterns, improved access to databases and other electronic resources, and the creation of a usability center to coordinate and conduct usability studies for the library and university."

Respondent 25

"Extended hours (user survey)."

"Adjusted workday reference desk hours (monitored statistics for a period)."

"Installation of a RFID security system (feedback from every assessment activity ever conducted, regardless of the topic)."

Respondent 26

"Extended library hours and new faculty communications vehicles as a result of LibQUAL+™ survey results."

Respondent 27

"Extended opening hours during exam periods."

"Participated in consortium virtual reference program. This evolved from an in-house service."

"Expansion of outreach activities: e.g., instruction program for faculty/students in academic departments' facilities; other liaison activities."

Respondent 28

"Extended service hours and expanded customer service training to better meet user needs."

"Increases in productivity and in cost savings."

"Improved design and content of Web site."

Respondent 29

"Facilities Renovation Plan."

"Public Services Competencies Development & Training."

"Improved access to online resources and assessment of library's Web site."

Respondent 30

"Focus group discussions with students have convinced us to do some renovations differently than originally conceived."

"Learning from survey results that good students were heavily unaware of liaison librarians, caused us to put more energy into reaching out to grad students."

"Decision to purchase or cancel subscriptions is partly based on usage stats."

Respondent 31

"Geology Library Renovation Survey: Questionnaire designed to inform decisions regarding study space needs of faculty and graduate students in the Geology and Geoscience departments. Results informed decisions regarding purchase of soft seating and study table size and location within the library. Also, graduate students repeatedly requested that lockers be made available, as they travel between two campuses. We were able to include lockers in the renovation, and were otherwise not intending to."

"Virtual Reference Assessment: Analyzed 5 years worth of chat and e-mail reference transactions. The following recommendations were acted upon, more are forthcoming: Cancel contract with Live Assistance Chat, switch to commercial IM (Meebo). Staffing of chat reference services was adjusted slightly, based on the distribution of subject areas of the questions asked. Staffing of CUL Suggestions (online suggestion box) was rotated, due to one division having staffed it for 5+ years. E-mail management system is on order to manage e-mail transactions from the public Web sites. Access Services and Reference Services are partnering. More assessment: Phase II of Virtual Reference Assessment will gather user input to inform further decisions about the structure of reference services (i.e., centralized vs. subject specific)."

"Document Delivery Benchmarking Survey: This survey was sent to peer institutions, inquiring about their current document delivery services and the planning processes for establishing these services. Cost was also addressed. Clarified the need for the Shipping Department to join Access Services in support of campus-wide delivery programs. Shipping joined Access in June 2007."

Respondent 32

"Help budget planning/justification."

"Data used in library marketing and communications."

Respondent 33

"Identification of priority services."

"Identification of facilities problems."

"Identification of quality of service to users."

Respondent 34

"Implementation of Customer Service program for Information Access staff."

"Shifting staff resources."

"Increase funding for Interlibrary Loan borrowing."

Respondent 35

"Improved service to users by increasing maintenance of equipment such as photocopy machines and increasing availability of public computers."

"Improved service to users by offering access to computer software such as Microsoft Office on public computers."

"Improved service to users by providing MyLibrary portal, RSS feeds, and integrating resources with campus courseware."

Respondent 36

"Improved signage to help students better navigate through the Main Library."

"Extended library hours and additional staff."

"Completely revamped library Web site to make navigation easier, resources more transparent, and functionality more intuitive."

Respondent 37

"Improved turnaround time for acquiring new books requested by users."

"Redefined collection development (disbanded department, reduced foreign language acquisitions, added copies of high use items, funded transition from print to e-journals, etc)."

"Improved speed and accuracy of shelving."

Respondent 38

"Improved procedures and processes for requesting and delivering print materials from storage facility."

"Revisions/improvements to libraries catalog and Web site."

"Program improvements to Libraries Open House based on participant feedback in exit survey."

Respondent 39

"In spring 2007 the library implemented longer hours as a result of information collected in a LibQUAL+™ survey earlier in the academic year."

"Two departmental libraries were closed based on information from surveys and statistics gathering."

"Based on feedback from library users, the library decided to open its electronic classroom as a computer lab when classes were not scheduled."

Respondent 40

"Increase in library open hours."

"Revision of the OPAC user interface."

"Additional training opportunities for library faculty and staff."

Respondent 41

"Intensive customer service training as a result of LibQUAL+™ Affect of Service results and improvement in scores on next survey."

"Process improvement in circulation and shelving based on survey results changes procedures for checking in materials."

"Purchased library materials in specific subject areas based on results of interviews with faculty and focus group discussions with graduate students."

Respondent 42

"Justified improving endowed librarian salaries."

"Extended library service hours."

"Constructed additional storage units to shift a larger percentage of collections off campus (while increasing document delivery services)."

Respondent 43

"Leadership changes based on results of staff surveys."

"Collection activities based on results of faculty surveys."

Respondent 44

"Major e-Library redesign project (i.e., library Web site)."

"Increased emphasis, in the library's instructional programs, on evaluating information resources, and distinguishing between trustworthy and untrustworthy information."

"Very specific improvements to physical facilities (e.g., improved lighting in stairwells, etc.)"

Respondent 45

"Move to electronic only format for most scholarly journals."

"Creation of regional library service."

"Develop prototype data curation systems that support needs across a range of disciplines: Virtual Observatory; Roman de la Rose; and Dry Valleys in Antarctica."

Respondent 46

"None."

"None."

"None."

Respondent 47

"Number of user stations increased due to accreditation recommendation and later assessment."

"1995 user survey influenced design/equipment in building expansion project."

"Collection decisions are made based on usability in many cases."

Respondent 48

"Re/allocation of resources (including changes to staffing numbers and patterns)."

"User-oriented redesign of library Web site."

"Increase in outreach efforts by librarians."

Respondent 49

"Replaced online catalog system with a new one."

"Started online reference services."

"Initiated self-service checkout and other self-service activities; improved photocopying by buying new equipment and making it easier to debit charges for copies. "

Respondent 50

"Revised building hours."

Respondent 51

"Revision of library's public Web site."

"Revision of reference staffing needs."

"Revision of digital library interface."

Respondent 52

"Routinely make adjustments to library subscriptions based on use statistics."

"Reconfigured the libraries' Web site based on user surveys and focus groups."

"Added positions in a liaison area based on interviews with colleges' administrators."

Respondent 53

"Selective reduction in journal subscriptions based on analysis of usage."

"Redesign of library Web site resulting from usability studies."

"Expansion of laptop loan service based on analysis of usage statistics."

Respondent 54

"Service quality improvements through better training and related goal setting."

"Process improvements in circulation and re-shelving."

"Adding more computers and software to meet user requests."

Respondent 55

"Significant modifications to library Web site design and functions."

"Changes in collection development budgeting process, also extensive modifications to quality control in and management of stack operations."

"Awarding of grants to certain discipline-based library programs."

Respondent 56

"Space improvements."

"Web and catalog redesigns."

"Emphasis on instructional services."

Respondent 57

"The library has made considerable facilities improvements based on LibQUAL+™ data and focus groups. The creation of two Commons areas has been greatly informed and facilitated by the assessment that has been done."

"The library's Web site modifications and improvements have been informed by LibQUAL+™, an outside consultant, and in-house usability testing that we've done over the last two years."

"The library periodically (most recent are 2000, 2004, 2006) does a Serials Review using a variety of factors including price, inflation rates, local use, availability at other local libraries or electronically through consortia agreements, ISI Impact Factors and ranking in relevant disciplines. Faculty input is an important facet of this process. This assessment is done in order to provide the best collections and resources to meet the needs our constituents."

Respondent 58

"The library introduced staff recognition programs to address workplace concerns identified in the Library Employee Opinion Survey."

"The library's Web page was refined/improved based on input from usability studies and focus groups."

"The library's ILL service was modified to remove barriers to access based on user survey."

Respondent 59

"The redesign of the libraries' Web site and the improvement of the OPAC."

"The closing and remodeling of branch libraries."

"The expansion of libraries hours."

Respondent 60

"Updated and improved library services."

"Increase staff training."

Respondent 61

"Upgrading of photocopy and printing facilities."

"Increased funding for technology upgrades."

"Increase funding for collections, both print and electronic."

Respondent 62

"We increased hours for the main library from midnight to 2 am during academic terms."

"We created an electronic-preferred journal policy."

"We have reorganized some library departments as a result of assessment efforts."

Respondent 63

"Web pages have been adjusted based on testing and feedback from students, faculty, and staff."

"Access policies have been changed based on feedback from patrons."

"Reserves policies and structure have been changed because of faculty needs and changes in technology."

Respondent 64

"Worked with consultants to redesign signage in response to data from a wayfinding study showing that previous finding aids were ineffective."

"Redesigned library carrels and chose new seating based on user surveys responding to model designs and on focus groups on user preferences for study areas."

"Redesigned library home pages based on user surveys/interviews, Web logs showing heavily used resources, and user feedback on prototypes."

## PROFESSIONAL DEVELOPMENT

### 18. Does your library provide assessment training for library staff? N=68

| | | |
|---|---|---|
| Yes, training is provided by the library | 19 | 28% |
| Yes, support is given for training provided by our parent institution | 22 | 32% |
| Yes, support is given for training provided outside of our institution | 42 | 62% |
| No, there is no particular training provided | 20 | 29% |

### 19. If training is provided by the library, what kinds of topics are covered? Check all that apply. N=24

| | | |
|---|---|---|
| Assessment methods | 14 | 58% |
| Basic statistics | 11 | 46% |
| Survey construction | 11 | 46% |
| Value of assessment | 10 | 42% |
| Data analysis | 9 | 38% |
| Data presentation | 7 | 29% |
| Sampling techniques | 6 | 25% |

| | | |
|---|---|---|
| Report writing | 3 | 13% |
| Other (please specify) | 7 | 29% |

"COGNOS PowerPlay and Impromptu, which are part of the COGNOS business intelligence software suite."

"Data collection, data entry, qualitative software usage (once)."

"Focus group techniques."

"Focus groups, qualitative software training (NVIVO), Excel."

"Human subjects regulations."

"University's Planning & Institutional Research department provides consultation/advice services. Other training programs are in preparation, e.g., using the LibQUAL+™ survey process as a case study in developing broader expertise."

"We are just beginning to provide training to the Assessment team."

20. For each of the following assessment-related professional development events that assessment staff have attended, please indicate whether they would or would not recommend the event to others as a good way to learn and network about assessment. Check "Have Not Attended" if no assessment staff have participated in an event. Check one category in each row. N=67

| | N | Would Recommend | Would Not Recommend | Have Not Attended |
|---|---|---|---|---|
| Library Assessment Conference (e.g., Charlottesville 2006) | 66 | 39 | — | 27 |
| LibQUAL+® training sessions | 66 | 54 | 1 | 11 |
| Northumbria International Conferences on Performance Measurement in Libraries | 65 | 10 | 1 | 54 |
| Evidenced-Based Library and Information Practice Conference | 64 | 9 | 2 | 53 |
| ARL assessment-related meetings | 64 | 56 | — | 8 |
| ALA/LAMA sessions/discussion groups on assessment | 61 | 34 | 1 | 26 |
| ALA/ACRL sessions/discussion groups on assessment | 61 | 37 | 3 | 21 |
| Other | 22 | 14 | — | 8 |

If you selected "Other" above, please specify which other assessment-related professional development event(s) assessment staff have attended. N=14

"ALA LRRT programs at ALA (depending on topic)."

"ALA/ORS sessions. LITA Preconference (and sessions? - usability). NFAIS (online usage statistics)."

"ARL online course, OLA 3-day special course."

"CARL meeting/workshop 2006."

"Creating an Environment of Continuous Assessment: Practical Approaches for Academic Libraries (OCLC Western Workshop)."

"I would also recommend ARL Assessment 'Boot camp,' New Orleans 2007. Re: Negative response for LibQUAL+™ Training Sessions: I attended one not sponsored by ARL, I believe it was NYLINK, and it was not helpful unless you've really not had any experience or knowledge of LibQUAL+™. The presenters did not seem to have mastered the subject matter."

"In the Canadian/Ontario context: sessions at Ontario Library Association Super Conference and the Canadian Library Association Conference. Education Institute (OLA): audio conference sessions on research policies, issues, and methodologies. Also, ARL's online course: Measuring Library Service Quality with Danuta Nitecki and Toni Olshen."

"LibQUAL+™ Canada 2007 Workshop."

"Living the Future conferences (University of Arizona & ARL)."

"Meetings of local institutions conducting assessment undertaken as a follow-up to ARL Assessment Conference."

"Non-library national/regional assessment conferences and workshops."

"Sessions presented as part of the annual Ontario Library Association conference."

"State-wide assessment workshop for academic librarians."

"Web-based seminars and conferences."

21. **Please describe any professional development needs that assessment staff at your library have that are not being met by the events above. N=15**

"Basic statistics, data analysis, data presentation, survey construction, sampling techniques, focus group administration."

"Basic survey and focus group techniques, data analysis techniques, best practices on pre and post assessment communications."

"Basics of understanding data and data analysis."

"Classes in statistics and survey design."

"General overview of statistical analysis of data."

"I would greatly appreciate a listserv for assessment staff for the benefit of information sharing, communicating, partnering, and peer group support."

"In-house training on statistical methods, specific tools, e.g., atlas .ti, SPSS, etc."

"More practical instruction on how to formulate survey and interview questions. There's a lot of discussion about do's and don'ts, but no opportunity to learn in a collaborative, hands on environment. I'd like to have someone critique the questions I write."

"One of the primary challenges related to assessment is educating staff about its value and the need for it in the current climate. Many staff view it as an intrusion and a threat. Changing the culture is very challenging."

"Only Northumbria and the Library Assessment Conference provide consistent educational and networking opportunities. Northumbria is often impossible to attend; LAC has only had one conference. The various meetings at ALA (LAMA, ACRL, ARL) are scattered and uncoordinated, sometimes even scheduled over each other. There are very few _sessions_ on assessment. There seems to be no one place that people doing assessment (in any size library) can turn for support and education."

"Practical level assessment training for staff at all levels of the library. When you don't have an expert on staff where do you begin with assessment."

"Survey writing skills."

"Tailored training for implementing learned skills into institutional reality. Public presentation of results."

"The Assessment Officer intends to enroll in statistics and research courses in our College of Education to gain a more thorough grounding in statistics and assessment methodology."

"We had consultants come in for two-day session but might be useful to have extensive, affordable consulting that helped derive a plan when staffing is limited to do assessment."

## CULTURE OF ASSESSMENT AT YOUR LIBRARY

22. Please rate your agreement with the following statements on a scale of 1 to 5 where 1 is strongly disagree and 5 is strongly agree. N=68

(Adapted from "Culture of Assessment I.Q. (Institutional Quotient)" by Betsy Wilson, University of Washington.)

| | N | 1 Strongly Disagree | 2 | 3 | 4 | 5 Strongly Agree |
|---|---|---|---|---|---|---|
| Assessment is evident in our library planning documents such as the strategic plan | 67 | 1 | 8 | 8 | 20 | 30 |
| Library administrators are committed to supporting assessment | 68 | 1 | 5 | 8 | 25 | 29 |
| Assessment results are used to improve my library | 68 | 2 | 3 | 11 | 28 | 24 |
| Assessment is a library priority | 68 | 1 | 7 | 14 | 25 | 21 |
| My library evaluates its operations and programs for service quality | 68 | 1 | 5 | 13 | 28 | 21 |
| My library has local assessment resources and experts | 68 | 2 | 12 | 20 | 25 | 9 |
| Staff accepts responsibility for assessment activities | 68 | 3 | 8 | 36 | 17 | 4 |
| There is support and/or rewards for staff who engage in assessment activities | 67 | 6 | 13 | 19 | 25 | 4 |
| Staff have the necessary assessment expertise and skills | 68 | 2 | 23 | 30 | 10 | 3 |
| Staff development in assessment is adequate | 67 | 5 | 26 | 25 | 9 | 2 |

23. Does your library have assessment plans for departments/units or a library-wide assessment plan? N=67

| | | |
|---|---|---|
| Yes, the library has an assessment plan for every department/unit | 3 | 4% |
| Yes, the library has an assessment plan for some departments/units | 13 | 19% |
| Yes, the library has a library-wide assessment plan | 20 | 30% |
| No, the library has no assessment plan | 36 | 54% |

## Selected Comments from Respondents

"A library-wide assessment plan will be developed in the 2007–2008 fiscal year."

"An assessment plan/program is in preparation."

"As well, assessment is integral to, and included in all work described in the Libraries Workload Committee Report for Librarians/Archivists for the academic year 2007–2008 in the priority areas of Teaching and Learning, Research and Scholarship, Collaboration and Liaison, Collections, and Access."

"Balanced Scorecard metrics serve, in part, as our assessment plan."

"Development of an assessment plan may emerge from strategic plan."

"Just developing a plan for assessment."

"Many departments include assessment in their individual strategic plans."

"Responsibility for assessment activities was only formally assigned in the past two months, but creating a library-wide assessment plan is a top priority for the library in the coming year."

"The Assessment Librarian position has been in existence for approximately 3 months, however an assessment plan for the library system is under development, in addition to the assessment staff resource Web site. The developing assessment plan is a proposal for adoption by the library system."

"The Information Access and Delivery services department is currently working on an assessment project to inventory what we collect and assess if we are collecting the right data and determining what we not longer need to collect. The group charter also includes determining the 10 highest priority measures."

"The libraries-wide assessment plan occurs in the form of the annual report of the Office of Assessment & Planning."

"The library system is currently in the process of further defining and articulating its assessment efforts and plans. Work is in the early stages, and ongoing among a number of individuals and groups."

"The plan could be more 'formal.'"

"The writing of a library-wide assessment plan is a priority for the Assessment Officer."

"This is an area that we are currently developing in a committed way."

"We are developing a library-wide assessment plan at this time."

"We are working on an assessment plan tied to our strategic plan."

"While we have indicated a strong desire for evidence-based decision making, we are still in the process of putting in place the necessary components. We had initiated the LSAC to begin work when it suspended activity while we completed our strategic planning work when our new Dean arrived. We now have key success measures to develop and then we will determine the best way to move forward — with LSAC, a different committee structure and mandate, etc."

## ADDITIONAL COMMENTS

24. Please enter any additional information regarding assessment activities at your library that may assist the authors in accurately analyzing the results of this survey.

"Although primary responsibility for coordination and planning of assessment activities rests with a standing committee, it should be noted that in 2007 the libraries created a full-time assessment position, Library Data Officer, reporting to the Dean of Libraries. Although the position does not have overall responsibility for coordination of assessment activities, it is charged with several of the responsibilities enumerated under questions 8–12; specifically: analyzes, interprets, and reports on data collected in assessment activities; consults with staff on assessment methods and needs; coordinates collection of data across the library; coordinates the reporting/archiving of the library's statistical data; fills requests for library data; performs assessment activities; provides training on assessment topics; and submits external surveys."

"Assessment activities at the seven service locations at Western Libraries are highly distributed. Western Libraries' system-wide assessment activities are coordinated by the Assessment Librarian."

"At our library, we are approaching assessment in a systematic way by first defining the mission of the Research & Assessment Unit. We are contracting a consultant to provide basic assessment training to unit members. We are also conducting a feasibility study for a local data mart to house data we collect."

"Conducted LibQUAL+™ in 2003 and 2006 but have not analyzed the differences yet."

"From 1998–2003, a standing library-wide team was in place to conduct user assessment activities. Since 2004, separate assessment activities have taken place but without library-wide coordination. A library-wide ad hoc committee was established to administer, analyze, and report the 2005 LibQUAL+™ survey. That group disbanded after the survey report was completed."

"Many items that were mentioned in this survey are in development stages at our library: data mining and a public Web site for assessment are two major initiatives that are in early planning stages, but will be realized in the near future. As part of the process of developing the CUL Assessment Plan, we conducted a Culture of Assessment IQ Test; the answers provided here are based on that survey. We will conduct this survey again, on tri-annual basis, in line with our planning cycle. We also offer a series of Assessment Forums to invite colleagues into the libraries to present their experiences with assessment to staff, and hopefully inspire some projects."

"Our unit with primary responsibility of assessment was only formed in the past year. We've had a standing assessment group, the Library Assessment Group, which has been in place since 1992: ad hoc 1992–1996 and made ongoing in 1997. This group is made up of 9 members including the two members of the Office of Assessment & Planning and works with the Director of Assessment & Planning on assessment projects and activities."

"The library's current approach to assessment has evolved from an interest group that formed as a subgroup of the library's committee on reference in November 2003. This voluntary group soon expanded to include staff from all library units, and this group evolved from an interest group to a formal 'resource group' in May 2005. This voluntary committee was extremely useful in educating staff about the need for assessment, in generating new activities, and in developing expertise. However, the need for a coordinated and programmatic approach to assessment become evident, and the Library administration decided to assign responsibility for its assessment activities to an organizational unit in early 2007. An Assessment Director was hired in May 2007."

"This library has put extraordinary effort into building management information structures as necessary components of its assessment mandate. The principal is to empower staff broadly to conduct assessment and own the priority to assess. While the assessment function properly belongs to the full staff, the development of management information resources is the focus of a central department, which also designs and builds tools and provides a level of central coordination. Our goal is to build a scale-able, staff-driven, and user-focused activity of the enterprise."

"Though the Assessment Officer does and will continue to have primary responsibility for coordinating and planning assessment activities, the University Libraries intend to appoint an Assessment Committee to work with him on matters of assessment and statistics gathering/analysis. This committee will also be involved in writing the library-wide assessment plan."

"The library is just beginning this process. January/February 2007: first participated in LibQUAL+™ Survey. January 2007+ Assessment Librarian's position (job description) is under discussion."

"We have recently begun the process of moving to a more systematic and programmatic approach to assessment. Like many institutions, we have undertaken a great deal of ad hoc activity in the past and this is reflected in survey responses, but the contours of a formal assessment program are very much still emerging."

"We intend in the future to increase the use of our statistics (logbook, etc.). We also have a Statistics Committee that coordinates collection of data across the libraries."

# RESPONDING INSTITUTIONS

University at Albany, SUNY
University of Alberta
Boston College
Boston Public Library
Brigham Young University
University of British Columbia
Brown University
University at Buffalo, SUNY
University of California, Berkeley
University of California, Davis
University of California, Irvine
University of California, Los Angeles
University of California, San Diego
University of Chicago
University of Colorado at Boulder
Colorado State University
Columbia University
University of Connecticut
Cornell University
University of Delaware
Emory University
University of Florida
Florida State University
George Washington University
Georgia Institute of Technology
University of Guelph
University of Hawaii at Manoa
University of Houston
Howard University
University of Illinois at Chicago
University of Illinois at Urbana-Champaign
Indiana University Bloomington
University of Iowa
Iowa State University
Johns Hopkins University
University of Kansas
Kent State University
University of Kentucky
University of Louisville
University of Manitoba
University of Massachusetts, Amherst
University of Miami
Michigan State University
University of Minnesota
Université de Montréal
University of Nebraska–Lincoln
New York Public Library
University of North Carolina at Chapel Hill
North Carolina State University
Ohio State University
University of Oklahoma
Oklahoma State University
University of Oregon
University of Pennsylvania
Pennsylvania State University
Purdue University
Rutgers University
University of Saskatchewan
University of Southern California
Syracuse University
University of Tennessee
University of Texas at Austin
University of Utah
Vanderbilt University
University of Virginia
University of Washington
Washington State University
Washington University in St. Louis
University of Waterloo
Wayne State University
University of Western Ontario
University of Wisconsin–Madison
Yale University
York University

# REPRESENTATIVE DOCUMENTS

# Position Descriptions

**University of Alberta Library**
**Librarian Position Description**

## I. POSITION INFORMATION

Position: Assessment Librarian
Administrative Unit: Director's Office / Administration
Supervisor: Director, Library Services & Information Resources

## II. GENERAL STATEMENT OF RESPONSIBILITY

### i. PRIMARY JOB FUNCTION

The position supports service improvements to users of the Libraries by coordinating assessment activities, interpreting data and studies, making recommendations to the Director and communicating with staff and users.

### ii. NATURE AND SCOPE OF RESPONSIBILITY

**The A**ssessment Librarian plans and implements campus-wide survey instruments developed by the Association of Research Libraries and other credible bodies. She also plans and implements local assessment tools and activities, as appropriate. This may require attendance at relevant meetings and training events. She interprets the results and brings them together to make recommendations to the Director and, through the Director, to SAT [Senior Administrative Team], regarding service improvements within the Libraries.

Survey data is integrated with other initiatives such as usability studies, focus groups, and computer-based management information and statistics to build and maintain a body of knowledge on user preferences and behaviors. Data is organized and stored to create trend lines over time to support library assessment efforts within the University Libraries and to form the basis of an ongoing, sustainable assessment program.

Creation of recommendations to SAT may involve research into best practices in the literature, and personal contact to establish practices at comparable institutions. As appropriate, recommendations adopted by SAT may result in the establishment of working groups to deal with specific improvements, and the incumbent may be asked to serve on such groups to provide evidence-based guidance regarding the implementation of changes.

The Assessment Librarian helps to foster a culture of assessment within the Libraries by developing expertise and understanding of assessment measures and techniques and sharing these with library staff. She provides coordination and support, as appropriate, for assessment activities of other library staff.

Based on an understanding of user interests, the incumbent compiles the Library News on the Libraries' web site, identifies stories about the Library and its staff, and alerts the editors of Folio and Express News. This may include writing news stories and updating information for students and faculty. The incumbent is also responsible for communicating assessment activities and results to staff in a timely and appropriate manner.

**III. WORKING RELATIONSHIPS AND COMMUNICATION**

**(i) INTERNAL**

Communicates regularly with the Director, and members of the Senior Administrative Team regarding service improvements based on user feedback and behavior; communicates irregularly with all staff regarding surveys and their results.

**(ii) EXTERNAL**

Communicates regularly through listservs and in-person with ARL staff on specific surveys; communicates irregularly with the campus community, campus news media, student associations, and peer positions in other institutions.

**IV. COMMITTEE AND PROFESSIONAL INVOLVEMENT**

**(committee responsibilities REQUIRED by the position and the position held on each committee)**

The incumbent is expected to contribute to the Libraries' leadership role within the community of research libraries through presentations and publications regarding assessment activities.

**V. SPECIFIC RESPONSIBILITIES**

**(five to seven statements which describe what your position is intended to accomplish and the duties involved)**

Identify opportunities for service improvements based on interpretation of survey results and other data / information sources

Make recommendations through the Director to the Senior Administrative Team

Communicate assessment plans and results to library staff

Provide coordination and support for assessment efforts conducted by other library staff

Create a body of knowledge on user preferences and behaviors

Participate in evidence-based implementation of service improvements

Contribute to campus knowledge of the Libraries by compiling Library News and writing for other campus media

**BRIGHAM YOUNG UNIVERSITY**
**STAFF/ADMINISTRATIVE JOB DESCRIPTION QUESTIONNAIRE**

**When submitting a signed job description, please provide us with:**
1. Department Organization Chart Including Employee Names
2. Electronic Copy of Job Description (via e-mail or computer disk)

**Compensation Dept Use Only:**
Points/Grade
☐ Exempt
☐ Non Exempt
Eval Date:

**POS#:**

**DATE: June 5, 2007**

**GENERAL INFORMATION:**

**SIGNATURES:**

**INCUMBENT:**

**JOB TITLE:** **Process Improvement Specialist** ______________________

**IMMEDIATE SUPERVISOR:** **, Associate University Librarian** ______________________
(Name and Title)

**DEPARTMENT HEAD:** **same as above** ______________________
(Name and Title)

**ORGANIZATIONAL UNIT:** **Harold B. Lee Library**

**JOB MISSION:**
Make a brief one or two sentence statement describing the primary purpose of this position.

Develop, implement, interpret and report on assessment, evaluation and usability studies that measure the effectiveness and efficiency of library operations. Collect and report data that supports decisions made by library department chairs, division heads, and various committees including the Library Administrative Council. Serve as Executive Secretary of the Consortium of Church Libraries & Archives (CCLA) to facilitate communications and to coordinate activities of the group.

**DIMENSIONS:**
List significant statistics upon which the position has impact.

- **Total annual income or operating budget:** $____________________
- **Total annual payroll:** $ $16,000
- **Other funds (if any) upon which the position has impact:** CCLA collections - $492,300; CCLA supplies/travel - $78,000
- **Personnel Supervised:**
  **-Direct** (reporting directly to this position):
  **Administrative**____ **Staff**____ **Student** 1 FTE **Non-students**___
  **-Indirect** (supervised by people reporting directly to this position):
  **Administrative**____ **Staff**____ **Student**____ **Non-students**_____
- **List several typical duties of those** directly supervised by this position:
- **Other Relevant Figures** (e.g., people served, volume of activity, etc.):

The work of this position is essential to decision-making in every area in the library including collection development, technical processing, public services and electronic resources. This position regularly monitors the library's collection expenditures budget which exceeds $6,000,000; data generated by this position heavily influences library-wide collection development programs and directions. The library circulates over 600,000 books each year, answers approximately 200,000 reference questions, teaches nearly 2,000 classes a year, and receives millions of page views to its website. This position designs and performs assessment studies that directly impact decisions relative to all these operations.

The Consortium of Church Libraries and Archives (CCLA) oversees cooperative activities of the Lee Library and the Hunter Law Library in Provo, libraries at BYU-Idaho, BYU-Hawaii, and LDS Business College, as well as the Church History & Archives, Church A/V Library, the Family History Library, and other libraries under the umbrella of the Church of Jesus Christ of Latter-day Saints. Jointly these libraries serve not only their local constituents, but also Church members worldwide. This position coordinates decision processes related to expenditure of the Consortium's collection budget of $492,300 and its operating budget of $78,000.

**ENVIRONMENT:**
Describe the context (setting) within which the position operates and provide any relevant information about influences on the position (e.g., regulations, economic conditions, technology, etc.), or special circumstances affecting the position (travel, night work, etc.).

This position helps the Lee Library and other libraries in the CCLA measure their performance and set new directions relative to collections and services. The data that is systematically analyzed by this position is used by the library in both long-term strategic planning and shorter-term operational design. The position must work with employees in every unit of the Lee Library and must be conversant with the mission, goals, functions, and operations of those libraries. Because the Lee Library offers services to distributed users, the position also works with librarians, faculty and students at the Hunter Law Library, BYU-Idaho, BYU-Hawaii, LDS Business College, Church History & Archives, Church A/V Library, the Family History Library and other partners in the CCLA. The broad geographic distribution of these libraries and the fact that each operates in its unique administrative context adds complexity to fulfilling responsibilities of this position.

A significant amount of high-level technology is used in planning, conducting, and reporting process improvement studies. The employee must be comfortable and self-sufficient in using technology. Because the working environment is strongly team-oriented, good interpersonal skills are necessary. The position must be able to make effective presentations to groups of librarians, students, faculty and administrators.

Process improvement studies will frequently focus on highly complex workflows. The position will typically be under pressure to complete studies and report findings on a tight schedule. Performing studies may require work during weekends and evenings.

This position is also responsible for managing, in cooperation with the Evaluation/Measurement Specialist for the Center for Teaching and Learning, the library's website usability center, which is the principle resource for coordinating and conducting website usability studies and related activities.

**LEVEL OF SKILL OR EXPERIENCE**
Specify special skills, abilities, experience, or knowledge needed to perform effectively in the position. Do not use your own qualifications.

Undergraduate degree in statistics or related field plus five years' experience required. Graduate degree in statistics or related field preferred.
Experience with current assessment research, practice and theory
Experience in methods of research and evaluation, statistics, testing, and measurement
Ability to analyze research and interpret results in support of decision-making
Experience using statistical software in research, program design, data analysis and reporting
Ability to work effectively with a wide variety of stakeholders including librarians, faculty, students and administrators
Experience using information technologies used in library services and higher education
Effective oral and written communication skills and experience making group presentations
Experience with process improvement and website usability studies desired
Experience with evaluating library operations desired
Experience with strategic planning desired
Strong organizational skills

**PROBLEMS AND CHALLENGES**
Note the nature and variety of the more challenging types of problems encountered or the kinds of problems to be solved.

Continually monitoring a dynamic environment to determine if library collections and services are meeting the needs of students and faculty

Designing and overseeing programs of assessment and evaluation that lead the library in rapidly adjusting to change

Communicating to librarians the importance of sound evaluation to effective planning and decision-making

Maintaining expertise in assessment techniques while also being conversant with trends in library service and operations

Generating reports and analyses in a short, intense time-frame when administrators call for them

Understanding highly complex library operations

Understanding library systems and being conversant with how to retrieve and analyze data from multiple systems.

Facilitating resolution of projects that require input from numerous librarians at institutions scattered geographically from Provo to Salt Lake to Hawaii to Idaho

Balancing the needs of the CCLA as a whole with the individual needs of local units

Manage the library website usability center in cooperation with the Center for Teaching and Learning

**DECISIONS MADE AND REFERRED**
Identify the kinds of decisions made by the position in matters relating to the **budgets, personnel, or other aspects of the job.** Does the incumbent have the authority to **hire or fire** without approval from the supervisor? Also identify what types of decisions must be **referred to the supervisor**.

Made:
- Design of process improvement studies and instruments
- Execution of approved assessments
- Determine best methods for assessing services across multiple libraries in the CCLA.
- Consult, coordinate and/or carry-out usability studies to improve the functionality of the library's website.

Referred:
- Process improvement studies needed
- Priority of assessment studies to be performed

**PRINCIPAL CONTACTS**
List any professional organizations the **position** participates in, any campus committees the **position** is a member of, or other significant internal or external contacts.

In HBLL:
- Library administrators and other library employees
- Library Administrative Council
- Library committees including Department Chairs, Coordinating Council, Reference Services, Collection Development, Information Literacy & Web Working Group
- Library project teams charged with process improvement or assessment (for example, Reference Statistics Team, Reference Services Team, Information Commons Team)

At BYU:
- Faculty Library Council
- University administrators
- BYU Office of Planning & Assessment
- BYU Center for Statistical Consultation and Collaborative Research
- Center for Teaching and Learning (particularly their Director and the Evaluation/Measurement Specialist)
- Faculty and students of BYU
- Librarians of Hunter Law Library

Beyond campus:
- Association of Research Libraries (primary HBLL Liaison to ARL on statistical issues)
- LibQUAL+® survey administrators (primary liaison and local administrator of national library service assessment projects)
- Association of College & Research Libraries
- Faculty, students and librarians of BYU-Hawaii, BYU-Idaho, LDS Business College, Church History & Archives, Family History Library, and other CCLA libraries
- Librarians and administrators from other academic research libraries

**MAJOR ACCOUNTABILITIES** include, but are not limited to the following:
**Rank in order of importance (most important first)** the major accountabilities of the position (4 to 7 accountabilities for a normal position). List those activities for which the "buck" stops here. List the **duties performed** to achieve that accountability and the **percentage of time spent on each accountability.** (Note: Identify what is done, rather than how it is done).

1. Measure the effectiveness and efficiency of library operations 35%
   a. Interact with faculty, students and administrators to gather information on the needs of library stakeholders
   b. Participate in identifying and prioritizing needed process improvement studies
   c. Design and lead process improvement and assessment studies of library operations
   d. Coordinate assessments that compare library programs in multiple CCLA entities
   e. Analyze, interpret and report results of studies
   f. Evaluate faculty and student success in using library collections and services.
   g. Evaluate faculty and student satisfaction with library collections and services.
2. Oversee quality of measurement data related to library collections and services 15%
   a. Develop and implement effective data collection procedures
   b. Compile, analyze, interpret and report data, as it relates to patterns of library services, collection growth and usage
   c. Report on library operations to various associations such as the Association of Research Libraries, the Association of College & Research Libraries, and the Utah Academic Library Consortium
3. Facilitate activities of the Consortium of Church Libraries & Archives (CCLA). 15%
   a. Participate on the management team of the consortium
   b. Evaluate and assess structure of the consortium and recommend solutions to enhance efficiency
   c. Advise the management team regarding processes to facilitate activities of the consortium
   d. Facilitate communication among members of the Consortium board and committees
   e. Oversee content and currency of the consortium's website
   f. Maintain an ongoing record of policies, actions and activities of the consortium
4. Participate in library-wide reviews and assessment studies 10%
   a. Lead library-related assessment activities and statistical analyses requested by university administrators relative to budget, accreditation, and other formal reviews conducted at the university
   b. Prepare presentations and written reports associated with library-wide reviews and studies
   c. Generate data associated with accreditation studies, on both a university-wide and college/department level

   d. Assist in designing and generating content for inclusion in reports to university administration and other entities as requested
5. Manage the library's website usability center in cooperation with the university's Center for Teaching & Learning — 10%
   a. Consult with library departments in the usability of their web pages
   b. Evaluate and assess functionality of the library website under the direction of the library's Web Working Group
6. Participate on library committees — 5%
   a. Ongoing membership on the Library Web Team.
   b. Involvement on other committees as assigned.
7. Train and mentor library employees in assessment and evaluation techniques. — 5%
   a. Act as a staff resource for library areas engaged in process improvement.
   b. Collaborate with other library employees to integrate evaluation and assessment principles into library functions and services.
8. Pursue continuing education in assessment, evaluation, process improvement and usability — 5%

<u>ADDITIONAL COMMENTS</u>
Please provide any other comments that will help make the job better understood.

http://www.lib.uci.edu/libraries/jobs/planning_director.html

Librarian Vacancies »

## DIRECTOR OF PLANNING, ASSESSMENT, AND RESEARCH

The University of California, Irvine Libraries are seeking a dynamic professional to join an enthusiastic staff in building a research library of innovation and excellence for a young and rapidly-growing university ranked nationally in the top universities. The successful candidate will work in a collaborative environment of evolving institutional expectations and manage change to assure continuous improvement of our high quality programs and services.

The **Director of Planning, Assessment, and Research** is a new position in the rapidly growing UCI Libraries, responsible for insuring that excellent research, analysis, planning, measurement, and reporting are integral parts of all programs and services. The Director develops and manages the Libraries' capability to conduct effective planning, and conducts ongoing and targeted assessments of evolving priorities and needs in order to provide high quality services and well-focused support for the highest academic priorities of the campus. The Director leads planning and analysis for the effective allocation of new resources and the reallocation of existing resources (including budgets, staffing, and space), to meet current needs and to anticipate future priorities. The position reports to the University Librarian, works closely with Division heads (AULs) and Department heads, and consults with an array of people and units across the Libraries and the university on a wide variety of issues.

The Director functions at the highest levels of conceptualization and analysis, while insuring the integration of results into managerial practice; and provides leadership and substantial contributions to the growth and continued success of the UCI Libraries.

### Specific Duties

#### Planning and research

- Leads the conceptualization, design, implementation, coordination, and management of strategic planning, special project management, report and proposal writing, and other administrative initiatives for the UCI Libraries system, in coordination with library managers.
- Creates a sustainable planning environment to enhance effectiveness and efficiency; manages the processes used for planning, resource allocation, and decision-making.
- Facilitates and supports the collaborative development of policies.
- Researches issues and provides background for the University Librarian on special projects.
- Drafts and edits position papers, presentations, articles, briefs, and other key documents.
- Works closely with campus administrators and participates in campus and UC initiatives. Represents the UCI Libraries on various University of California committees and ad hoc groups and in local, regional, university, and national arenas as appropriate.
- Serves on university and library committees, task forces, and teams as assigned.

#### Assessment, evaluation, and reporting

- Responsible for creating, conducting, coordinating, and supporting standing and ad hoc processes for assessing service effectiveness, using appropriate instruments such as user surveys, questionnaires, and focus group interviews.
- Serves as an internal consultant on assessment to all units to ensure the appropriate linkage of planning, budgeting, and assessment
- Works with departments to analyze and distribute data to support assessment, management, and planning

http://www.lib.uci.edu/libraries/jobs/planning_director.html

- Develops, implements, and manages program for gathering, analyzing, and reporting management information and statistics
- Establishes standards for and monitors all internal data elements
- Develops and maintains statistical analytical skills among staff.
- Working with managers, maintains statistics and compiles narrative and statistical reports on activities and trends for reports.
- Coordinates the UCI Libraries responses to ongoing surveys such as the ARL statistics, accreditation requests, University of California surveys; and other surveys and inquiries
- Provides direction for and coordinates use of management information systems and the campus data warehouse.

**Capital space planning**

- Manages the development and implementation of capital space plans for the UCI Libraries.
- Develops plans for capital and renovation projects
- Monitors progress on various facilities projects, and capital project budgets and expenditures
- Serves as the Libraries' representative and liaison to various University offices including Capital Planning.
- Works with the Director of Development to coordinate fundraising efforts for capital projects.

**Contracts and grants**

- Oversees and coordinates grant application process
- Monitors all UCI Libraries contracts and grants, ensures appropriate reporting and financial controls, and ensures compliance with policies of the campus and external funding agencies

**Required Qualifications**

- Extensive knowledge of and substantial experience in managing programs, projects, and operations within a complex organization.
- Flexibility to initiate and adapt to change, and manage competing priorities.
- Relevant experience in one or more higher education organizations.
- Superior analytical skills, creative and innovative problem-solving skills, and a strong commitment to service excellence.
- Mastery of statistics and numbers.
- Expertise in relevant software applications.
- Demonstrated oral communication skills with the ability to be an effective and active listener and facilitator.
- Demonstrated written communication skills with the ability to produce executive-level reports and presentations.
- Demonstrated outstanding interpersonal skills with the ability to accomplish objectives with sensitivity and diplomacy.
- Ability to work effectively with staff in a multicultural and diverse environment.
- Demonstrated superior organizational abilities including facilitation and consultation skills.
- Knowledge and understanding of key issues and trends in higher education.
- Bachelor degree in a relevant discipline.

**Preferred Qualifications**

http://www.lib.uci.edu/libraries/jobs/planning_director.html

Graduate degree in library or information science, business administration, or a relevant discipline.

- Relevant experience in an academic or research library.
- Experience with space planning and management of multi-year, large scale projects.
- Advanced knowledge of statistical theory.

**The Libraries**

The UCI Libraries are committed to innovation and excellence and are in a major period of growth and change. The Libraries consist of the Langson Library, the Science Library, the Gateway Study Center , and the Grunigen Medical Library. The Langson Library primarily serves the Schools of Humanities, Arts, Social Sciences, Social Ecology, the Graduate School of Management, the Department of Education, and Interdisciplinary Studies. The Science Library primarily serves the School of Medicine and the Schools of Biological Sciences, Physical Sciences, Engineering, and Information and Computer Sciences. The Grunigen Medical Library serves the UCI Medical Center , located in Orange , 12 miles from the main campus.

The UCI Libraries have a staff of 273 FTE and an organizational structure that includes the use of teams in conjunction with departments. The library collection consists of over 2.3 million volumes and over 25,000 current serial titles and an aggressively expanding electronic resources collection. The UCI Libraries are a member of the: Association of Research Libraries (ARL), California Digital Library (CDL), Coalition of Networked Information (CNI), Center for Research Libraries (CRL), Council on Library and Information Resources (CLIR), International Federation of Library Associations and Institutions (IFLA) and Scholarly Publishing & Academic Resources Coalition (SPARC).

**University of California , Irvine**

One of ten University of California campuses, the University of California, Irvine, is nestled in over 1,500 acres of coastal foothills, five miles from the Pacific Ocean, between San Diego and Los Angeles. Founded in 1965, UCI is among the fastest-growing University of California campuses, with more than 24,000 undergraduate and graduate students and about 1,400 faculty members. UCI has had an extraordinarily rapid rise to distinction in its first forty years, including membership in the Association of American Universities, ranking tenth among the nation's best public universities by *U.S. News and World Report* (also among the top fifty research universities), and three Nobel prizes in the last nine years.

Student enrollment is expected to reach 32,000 by 2010 accompanied by a proportional growth in faculty. Nearly 60% of UCI students identify themselves as Asian American, African-American, Chicano/Latino, or Native American. The University offers 40 doctoral programs in addition to the M.D. UCI's academic programs are ranked nationally in the top universities; several doctoral programs are ranked in the top ten.

**Salary:**

Commensurate with qualifications and experience anticipated to be from $62,000 to $90,000

**Deadline for Applications:**

Applications received by November 29, 2005 will receive first consideration, but applications will continue to be accepted until the position is filled.

**To Apply**:

Qualified applicants who wish to be considered for this position should send their letters of application, complete résumés, and the names, e-mail addresses, and phone numbers of three references to:

Lillian Gates, Library Human Resources
UC Irvine, P.O. Box 19557
Irvine, CA 92623-9557
e-mail: ljgates@uci.edu
confidential fax (949) 824-3270.

Electronic applications are preferred. Upon application, candidates should be in possession of proof of their legal right to employment in the U.S. In compliance with the Immigration Reform and Control Act of 1986, verification of legal right to work will be required between the time of final selection and hiring, and is absolutely essential in ultimately being hired.

This position description is listed on the UCI Libraries Web site at http://www.lib.uci.edu/libraries/jobs/libvac.html, with links to additional Web sites featuring campus and community information.

***The University of California, Irvine is an equal opportunity employer committed to excellence through diversity.***

◄ Back

Top of Page ▲

**UNIVERSITY OF MASSACHUSETTS**
**CAMPUS:** ***Amherst***

**JOB DESCRIPTION**

*OFFICIAL TITLE: This is the official title of the position.*

Librarian III – 50%

*FUNCTIONAL TITLE: This is the in-house title by which the position may be known. A functional title is usually a more descriptive title than the official title and may be required to identify very specific kinds of work. This title may be used in signing all correspondence.*

Assessment Librarian

*GENERAL STATEMENT OF DUTIES: Please provide a brief overview of the general functions of this position.* Specific *details of duties should be reserved for the* Examples of Duties *section.)*

Develop and coordinate an assessment program for the Library. Take the leadership role in identifying, and developing appropriate assessment approaches, measures and techniques. Work with staff to ensure a broad understanding of the assessment goals and the ability to work with assessment measures. Represent the library for assessment initiatives on a national and local level.

*SUPERVISION RECEIVED: Please indicate the title, but not the name, of the administrative employee or employees responsible for supervision or direction of work; describe the divergent extents of authority of each, indicating the degree, priorities, and relationships of the supervision or direction, which could range from close supervision to supervision with considerable freedom.*

Report to the Director of Libraries with substantial independent responsibility and without immediate supervision.

*SUPERVISION EXERCISED: Using descriptive non-numerical terms, identify the scope of supervision, training or direction exercised (i.e., whether the supervision is over a few employees, a small number of employees, a large number of employees, etc.); also, describe the degree of supervision, indicating whether close supervision or general direction is involved, and categorize the physical conditions under which the supervision is given, such as in a laboratory or an office. Supervision of student employees should not be included in this section, but may be listed under* Examples of Duties, *if applicable.*

Exercise working supervision over professional and classified staff members engaged in the performance of assigned duties related to assessment.

*EXAMPLES OF DUTIES: Please list and briefly describe several of the duties and responsibilities typically performed and assumed in this position. This list should not be restrictive but should be descriptive in such a manner as to provide concrete information representing examples of the actual work as well as the level of responsibility for the work being performed.*

1. Develop, coordinate and promote assessment programs for the Library. Take the leadership role in identifying, defining and developing appropriate assessment approaches, measures and techniques.
2. Analyze, interpret, and publicize the results of assessment and evaluation activities.
3. Monitor and update the data gathering, statistics profile, and assessment measures used by the University Library. Maintain awareness of the administrative expectations on campus for measurement and assessment by the Library.
4. Represent the library for assessment initiatives on a national, regional, and local level.
5. Form and work with a staff committee to implement assessment initiatives.
6. Encourage and support assessment activities and staff engaged in assessment throughout the library.
7. Use both quantitative and qualitative measures according to established standards.

8. Participate in the pursuit and preparation of strategic opportunities including funding.
9. Perform other related duties as assigned.

*QUALIFICATIONS: Please indicate in a general way the knowledge, abilities, skills, education and experience necessary for any individual to assume this position. It is not the objective of this section to list any one person's specific personal traits and training. It is important to indicate, also, what degree of competence would be required (i.e., considerable education, extensive experience, working knowledge, etc.) to perform the duties and assume the responsibilities typical of this position.*

1. Master's degree in library science from an American Library Association-accredited library and information studies program.
2. Minimum of five years of professional library experience in this or another academic, research or specialized library which includes experience with assessment and analysis.
3. Demonstrated experience in assessment via quantitative and/or qualitative means required. Experience in manipulating raw data in order to analyze, customize, and format appropriate reports preferred.
4. Experience in planning, organizing and supervising the work of others and planning and organizing projects.
5. Demonstrated high level of written and oral communication skills and the ability to communicate clearly in presentations.
6. Ability to interact effectively and work productively, cooperatively and collaboratively with a variety of individuals and groups in a complex and rapidly changing environment.

***OFFICIAL POSITION CERTIFICATION***

**This is a complete and accurate description of this position.**

**Date** **Signature--Staff Member**

**Date** **Signature--Director of Libraries**

Profjd/LewellenLH-50%

University of Southern California
USC LIBRARIES

**Internal Position Addendum (20%)**
Assessment Coordinator

---

Data Gathering (7%)
Coordinate the gathering of monthly and annual library statistics from all data maintainers: establish process workflow, define data points, develop forms and templates, set accuracy standards, and train maintainers. Manage user surveys (such as *LibQual*); tabulate qualitative feedback (such as *Dear Library* and focus group feedback); and assist library personnel in developing local surveys as needed.

External Reporting (5%)
Coordinate responses to external requests for USC Libraries data within stated deadlines: including ARL annual surveys, institutional research surveys, and higher-education surveys facilitated through the USC Provost's office.

Internal Reporting (3%)
Develop quarterly and annual reports on library statistics and make available on library intranet. Respond to internal requests within stated deadline.

Assessment Team Leadership (2%)
Coordinate assessment assignments, roles, and responsibilities; develop assessment policies, goals, and strategies; manage assessment-related projects; and seek training opportunities in assessment for library personnel.

Data Management (2%)
Maintain authoritative data set for USC Libraries statistics and feedback. Manage assessment intranet site by posting, organizing, and updating documents.

Communication (1%)
Establish clear methods of communication in order to make library data available to all USC Libraries personnel, and to encourage their feedback. Disseminate data on campus and externally by managing a public web page; and engage library user-survey participants in follow up discussions.

# GENERAL FACULTY
# POSITION DESCRIPTION
# UNIVERSITY OF VIRGINIA

## PART I: ORGANIZATIONAL INFORMATION

| | |
|---|---|
| POSITION NUMBER | FP 558 |
| WORKING TITLE | Director of Management Information Services |
| ORGANIZATION | University Library |

## PART II: POSITION INFORMATION

| PERCENT OF TOTAL WORKING TIME | Describe fully and clearly the most important work first, then the next most important in a separate paragraph, and so on. Make the descriptions definite and detailed to give a clear picture of the work. It is particularly important that the supervision received or exercised be clearly explained. It is also important that such terms as "check", "handle", responsible for", "assist" be explained. List any equipment operated or special skill used. Estimate working time devoted to each kind of work described. |
|---|---|
| 50% | Manage the MIS unit:<br>• Provide leadership and vision for the MIS unit<br>• Administer the day-to-day operations of the unit<br>• Supervise, coach and mentor one Associate Director, and one Library Technical Assistant II.<br>• Develop, implement and oversee the compilation of data for the Library's major surveys<br>• Monitor and assist in the development of Balanced Scorecard metrics and measurements<br>• Oversee the development and reporting of the Library's Annual statistics – For local and national compilation<br>• Facilitate focus group activity, compile and analyze data<br>• Align UVA statistics with VIVA, ARL, and Project Counter<br>• Oversee the development and enhancement of an online data farm.<br>• Participate in all activities to create an environment of assessment such as presentations, meetings with departments, and training. |
| 20% | Serve on the Collections Steering Group and other library groups.<br>• Serve as the assessment and data resource person for the Collections Steering Group.<br>• Work to increase utilization of assessment data in collections decision making.<br>• See that collections budget expertise is transferred to the new budget manager.<br>• Serve on the usability team; work with the chair and other members of the team to see that a program of effective usability testing is institutionalized within the Library.<br>• Serve on the collections management group. Provide needed data, and context, regarding collections size, shelving needs, and budget<br>• Serve on the Promotions Review Board. |
| 15% | Compile, Report, Analyze and Manipulate MIS data for Administrative Reporting<br>• Prepare reports as needed for library administration and other units<br>• Present data and analysis at internal University Library events<br>• Answer queries from other libraries and organizations on behalf of UVA<br>• Collect, analyze, and report data from ARL and other sources, that will be useful in explaining and supporting the role of the Library within the University. |

Page Å

15% Professional Development & Service Activities

- Participate as the Visiting Program Officer for the Association of Research Libraries
- Attend professional meetings
- Present at professional conferences
- Publish articles in professional journals
- Serve as a mentor for other UVA librarians who wish to become professionally active
- Act as an Assessment Consultant at other libraries.
- Meet with colleagues locally and nationally regarding library assessment.

## PART III: QUALIFICATIONS

*Preferred knowledge, skills, and abilities for an individual performing this position:*
Ability to communicate effectively with faculty, staff, and students.
Ability to develop tools for collecting survey and statistical data.
Ability to collect, analyze, and report on statistical data.
Organizational and managerial skills.
Teamwork skills along with ability to work independently.
Commitment to participation in professional activities.
Commitment to customer service.

*Special licenses, registration, or certification:*

*Education or training (cite major area of study):*
MLS degree preferred or Master's degree in a related field required.

*Level and type of experience:*
Administrative experience in an academic library or similar institution.
Experience in collecting and reporting statistical information.

## PART IV: SIGNATURES

Signature Print Name Date

Current Incumbent

Individual Who Will Sign Performance Evaluation

Page Å

**UNIVERSITY OF WASHINGTON LIBRARIES**

**POSITION DESCRIPTION**

**POSITION TITLE:** Director of Assessment and Planning

February 2006

## GENERAL DESCIPTION

The Director of Assessment and Planning provides leadership and vision for the Libraries assessment and planning activities and is responsible for ensuring that assessment, measurement, planning and analysis are integral parts of Libraries programs and services. The Director oversees assessment efforts within the University Libraries; serves as an internal consultant for assessment activities conducted by other library staff; works with Libraries areas to analyze and report assessment data; represents the Libraries in campus, regional and national assessment efforts; evaluates the effectiveness of library assessment efforts; and recommends ways to strengthen the Libraries assessment and measurement program.

The Director creates and maintains a sustainable planning environment and provides information, analysis and reports to support Libraries planning and management activities. The Director establishes, manages, and provides access to management information; coordinates the ARL statistics and handles other centrally reported data and requests such as ARL SPEC kits and IPEDS surveys.

The Director oversees the work of and provides guidance to the Assessment and Planning staff; serves as Chair of the Library Assessment Group; and reports to the Senior Associate Dean of Libraries. Serve as an ad hoc member of the Libraries Cabinet and as an ex-officio member of the Libraries Council, the Strategic Planning Team,, the Diversity Committee and other groups as appropriate.

## SPECIFIC RESPONSIBILITIES

### Assessment

1. Direct assessment efforts within the University Libraries. Initiate assessment activities and provide consultation for assessment work done by other library staff or units. Maintain awareness of other related assessment efforts within the Libraries, University and externally. Serve as chair of the Library Assessment Group.

2. Analyze assessment-related data and communicate assessment activities and results to appropriate individuals and groups, including library staff and the UW campus community.

3. Develop and maintain expertise in assessment methods, techniques and best practices. Establish training as necessary for library staff on use of appropriate assessment tools and methods in collaboration with the Director of Organizational Development and Training.

4. Evaluate effectiveness of library assessment activities on a regular basis and make recommendations on ways to strengthen assessment work, including support needed. Foster and enhance the Libraries culture of assessment.

5. Respond to, collaborate with and participate as appropriate in other campus, regional and national assessment-related efforts.

**Planning**

1. Coordinate the ARL statistics and other UW Libraries responses to data requests and surveys such as the ARL SPEC kits and IPEDS surveys. Serve as the Libraries liaison to these groups and organizations.

2. Develop, implement and manage program for gathering, analyzing and reporting management information and statistics. Make relevant management information data accessible to Libraries staff and other appropriate parties.

3. Create and maintain a sustainable planning and analysis environment that is responsive to Libraries and University needs. Provide data analysis and reports to support effective Libraries management, planning, programs and services.

4. Work with the Strategic Planning Team and other Libraries groups in developing Strategic Plan performance measures and benchmarks, monitoring progress toward achievement of goals and objectives, and adjusting measures as appropriate.

**Assessment and Planning**

1. Supervise and guide the work of Assessment and Planning staff.

2. Undertake special projects as needed and assigned. Assume other responsibilities as assigned; perform other duties as required.

## POSITION DESCRIPTION

**Official Title/Title Code** Manager/1162

**Position Number** 39280

**Appointment Status** Administrative/Professional; Permanent; 12 months; 1.0 FTE

**Organization and Location** Washington State University Libraries, Pullman Campus/Holland 472

**Working Title** Library Data Officer

**Basic Function** This position provides expertise and advice to the Libraries in collecting, analyzing, and managing operational statistical data, and assists the Libraries in assessing the quality of its services.

**Reports to** Dean of Libraries

**Supervisory Responsibilities** N/A

**Duties and Responsibilities** (All listed duties are essential functions.)

Administrative Professionals are expected to communicate well; maintain pleasant, courteous and cooperative relationships; display a professional manner in demeanor and language; and show courteous and effective behavior in meetings.

**Data collection and analysis:**

- Compile and deliver statistical information in support of assessment and accreditation of University programs, and for surveys and reports.
- Serve as a primary resource in creating a culture of assessment in the Libraries.
- Work with library units and working groups to determine their statistical needs.
- Develop statistical databases and generate regular and special reports for library units.
- Collect annual statistics from library units. Prepare annual statistics for the Association of Research Libraries, National Center for Education Statistics and other bodies and projects. Maintain collected statistics in appropriate formats for preserving and sharing the information. Maintain Libraries' statistical web page.
- Train others on reporting software.
- Serve as a resource for library units and groups desiring assistance in identifying, gathering, and manipulating statistical information needed to assess their services.
- Coordinate statistical initiatives involving multiple library units or groups.
- Bring issues that are an outgrowth of identified statistical trends to library units.
- Assist units in collection and interpretation of statistics to further the goal of providing increasingly efficient and effective service.
- Coordinate ongoing statistics collection and reporting activities for collection development and management, such as journal use statistics and interlibrary loan statistics.
- Work with Library Systems to coordinate the compilation and analysis of usage

statistics for Libraries' web sites.
- Serve as *ex officio* member of the Assessment Working Group.
- Serve as *ex officio* member of the Mid Managers Working Group.
- Serve as Libraries' liaison with the WSU Office of Institutional Research.
- Serves as official WSU Libraries' representative to institutional, regional and national organizations at the request of the unit head, Assistant Deans, or Dean.

**KNOWLEDGE SKILLS AND ABILITIES**

- Knowledge of and skills in the use of computers.
- Ability to appropriately handle confidential information.
- Ability to prioritize and schedule assignments for oneself and others.
- Ability to adapt to departmental, University, and community needs as they develop and change.
- Ability to prepare and present ideas and information clearly and concisely in both written and oral form.
- Ability to gather information from a variety of sources and compile conclusions into a clear, concise, comprehensive document.
- Commitment to Diversity – an understanding and appreciation of the benefits of a diverse workplace and the knowledge of how to shape processes and procedures that reflect and enhance the diversity of the WSU communities.

**SIGNATURES BLOCK**

This position description describes the essential functions and qualifications for the position. This document does not exclude an opportunity for modifications consistent with providing reasonable accommodation. This position description is not a contract. Your signature indicates that you have read this position description and understand the essential functions and qualifications of the position.

Employee Date

Supervisor Date

pd39280.doc (Rev. 1.31.07)

# Assessment Charges and Mission Statements

http://libweb.lib.buffalo.edu/sw/committees/sms/about-sms.htm

staffWEB
University at Buffalo Libraries

# STATISTICS, METRICS & SURVEY DEVELOPMENT COMMITTEE

## about the Statistics Committee

### committee charge

The Statistics, Metrics and Survey Development Committee is responsible for reviewing and assessing the responses of the University Libraries or units to all external statistical and qualitative surveys of our resources, services, personnel and expenditures as well as all surveys we distribute to users or supporters of the University Libraries.

Specific responsibilities include, but are not limited to:

1. Review and oversee the completion of statistical and qualitative surveys of University Libraries resources, services, personnel and expenditures for international, national, state and regional associations and government agencies (e.g.: ARL, AAHSL, US Dept. of Education, SUNY, etc.) using appropriate library systems (e.g., ALEPH) that generate relevant statistical data. Charge task forces, as needed, to oversee the administration of major survey projects, such as LibQual+
2. Periodically review, compare and analyze the data collected for the various statistical and qualitative surveys (included in #1 above) and provide summary reports for review and discussion by the UL Directors Council, University administrators, faculty, and library faculty. Develop specifications for managing and representing the data that has been collected for future use.
3. Assemble and provide any metrics needed to assist the Associate Vice President for University Libraries in making budget requests or developing strategic planning documents.
4. Complete an inventory of all current and past statistical and qualitative surveys (included in #1 above) for which the University Libraries or individual library units have provided data and develop a spreadsheet or database with detailed information about each survey (e.g.: sponsoring organization, subject scope, date(s) when data were last provided, where the data is stored, the person responsible for compiling the UB data, etc.). Make this spreadsheet or database accessible to all members of the UL Directors Council, and as appropriate to all UL staff, on a central libraries server. Coordinate the archiving of completed surveys.
5. Serve as a clearinghouse for the review and inventory of all new survey instruments aimed at users or potential supporters of the UB University Libraries or of individual library units.
6. Maintain liaison with the major organizations developing and maintaining statistical and qualitative survey data about research libraries and alert the Directors Council to the need for UB representation at important meetings concerning these surveys. Review the current library literature as well as relevant Websites and maintain a bibliography of relevant current articles and information resources about trends and issues surrounding research library statistics, metrics and surveys.

### conducting business

#### Membership

Members of the committee are appointed by the Office of the Associate Vice President with approval by the University Libraries Directors Council for an initial term of three years.

- View the current membership roster.

#### Meetings

The committee communicates primarily by e-mail and calls meetings only as needed.

Content provider: Gary Byrd
Comments: lib-staffweb@buffalo.edu
Last update: 26 March, 2007

## University of Connecticut Libraries

USER Team

*Charge*
The USER Team is charged with coordinating library assessment and survey efforts for the University Libraries. The team designs, develops, conducts, and analyzes user surveys, coordinates the Libraries' participation in national user survey efforts, assists library staff in developing, conducting, and analyzing internal surveys, and archives survey data. The team develops reports for library staff that document survey findings for use in developing practical application of the data to the development of system-wide library policies, services, and collections. The team is also responsible for performing appropriate comparative analyses and for presenting data and statistical reports to the Libraries' Leadership Council (LC). Data and reports from USER team surveys are currently archived on the staff LAN and in the Dodd Research Center.

The team is also charged with coordinating collection of the Libraries' metrics and archiving this data for comparative and decision making purposes.

The team is expected to conduct or coordinate:

- Library user surveys every three years
- LibQUAL+ survey every three years or as appropriate
- Library assessment activities, such as Library metrics, annually by functional area
- Review and approval of all internally developed surveys
- Archiving of all data generated by internal and external surveys
- Organizational assessment every three years
- Other data gathering efforts as needed and determined appropriate by Leadership Council, Directors, and/or library staff

*Communication*
The USER Team will work in coordination with Leadership Council, library staff, and library functional areas. The team also develops annual goals and mid-year goal assessments, works with library staff to develop data gathering tools as requested, and for keeping staff apprised of survey efforts and reports via team participation in open meetings as appropriate, email communications, and the Libraries' home page.

Surveys on the Libraries Web page: http://www.lib.uconn.edu/about/administration/surveys/

Documents on Alectrona at: http://alectrona.lib.uconn.edu/Teams/Documents.cfm?Dir=User

Documents on the LAN (restricted to USER Team members) at: \\staff\data\userteam

E-mail distribution list: USER

*Team Composition/Selection*
The team is comprised of approximately four staff volunteers who are selected for specific skills, such as desk-top publishing, quantitative and qualitative data analysis, report writing, library collections, services, and strategic initiatives, and knowledge of university systems that provide information about campus populations. Members serve for three years in order to provide team continuity through a three year survey cycle (i.e., faculty/undergraduate student/graduate student surveys). Individuals who wish to remain on the team at the end of their term will be given first preference. Library staff should consult with their area head prior to submitting their name as a volunteer for membership.

*Leadership Council Liaison:*
The Director of Collections Services serves as the team's Leadership Council liaison.

The Leadership Council liaison will:
- meet each June with the newly populated team to review all aspects of the charge, or once a year in the case of a continuing team to ensure team members understand Leadership Council's intentions in charging the team
- be available to the team
- act as a communication link between the team and Leadership Council
- encourage the team to follow good project and team management skills
- ask questions or make suggestions if the team needs assistance
- be sufficiently knowledgeable of the work of the team so that the liaison is prepared to answer questions regarding the team's work
- advocate for the team when necessary or appropriate

*Term of Service*
The team will have 4-5 team members; two permanent and two/three rotating team members, who serve a minimum term of three years.

*Team Leadership*
The team selects a team leader from among the team members for a two year term.

While the work of the team is the responsibility of the team as a whole, the team leader is responsible for making sure the work of the team is done and for seeing that the following duties are carried out:
- meetings are scheduled
- meeting agendas are created and distributed
- meetings are run effectively and efficiently
- see that new team members joining an existing team are mentored
- meeting decisions and action items are documented and archived on Alectrona
- regular updates are sent to all staff to keep them informed of the team's work
- a mid-year report is submitted to Leadership Council (accomplishments to-date, obstacles encountered, and any requests for resources or revisions to goals set)
- an annual report is written
- annual goals are established and reviewed on a regular basis
- appropriate data are gathered and reported
- may also contribute to team members' annual performance evaluations.

*Reporting Relationship*
The team reports to Leadership Council.

*Meetings*
The team meets monthly or as appropriate to accomplish team goals and responsibilities.

*Minutes*
Minutes are recorded for each meeting on a rotating basis unless otherwise determined.

*Resources*
The team identifies human and fiscal resources needed for team training, software, or to conduct print and/or online surveys, or to perform quantitative and qualitative data analysis. Resources are included in the team's annual goals and presented for consideration to Leadership Council as part of annual and mid-year budget discussions.

**Charge for PSA Research and Assessment Unit**
*Approved by LMT on Feb. 7, 2007*

**Mission**

The Research and Assessment Unit (RAU) assembles and assesses data to report on Library performance and to provide evidence and context for Library priority-setting and decision making.

**Guiding Principles**

The Research and Assessment Unit

- Relies on functional experts in its work and trusts the wisdom and contribution of its colleagues;
- Strives to stay data-neutral in its research and analysis;
- Commits its resources to high-priority library projects, regardless of whether Public Services and Assessment (PSA) "owns" the project or not;
- Produces user-friendly products with trustworthy quality and timeliness;
- Recommends but does not enforce innovative changes for functional units

**Organization**

RAU resides within the Public Services and Assessment Division. It receives directional guidance from the Library Management Team (LMT) or through the Senior Associate University Librarians. RAU collaborates with other library and University units and groups, such as PSEC, Technical Service Senior Management Team, ITEC, CDExec, Library Communications, Administrative Operations, and the University Office of Institutional Research and Planning, to execute its assignments.

**Scope**

RAU's primary responsibilities fall in three categories:

1. It commissions data from library units. It organizes, analyzes, and reports on CUL performance to support reporting requirements, process improvement, and decision making.
   - It defines CUL's responses to data requested by ARL, ACRL, the NationalCenter for Education, and other external entities.
   - It communicates the definitions to appropriate functional and organizational units within the library.
   - It assembles and interprets data reported by the units and submits them on behalf of CUL to requesting entities.
   - It produces the CUL Annual Statistics Report and redeploys collected data in support of CUL activities and public relations when called upon.

2. It maintains and improves a shared library data repository to support library-wide assessment.
   - It partners with experts in other library units to build a central data repository that facilitates the storage, preservation, and use of library performance data in both qualitative and quantitative forms.
   - It repurposes assessment data, such as survey results, queries, collection analysis, to support on-going self-examination of library operations, collections, and services.
   - It strives to make data submission and use easier for both data producers and consumers.

3. It conducts high-priority research and assessment projects to support
   - Evidence-based decision making
   - Usability analysis
   - Innovation

In this role, RAU focuses on the usefulness and impact of library services. It assists library project teams and functions both as an "intelligence unit" that conducts environmental scans in the project area, spots patterns and trends, and suggests innovative changes; and also as a "consultant" that recommends techniques for assessment, and participates in assessment instrument design, administration, analysis and reporting as requested by the project. RAU relies on the project team for its expert knowledge and works closely with designate resources from the project team when executing its assignments.

Relying on a network of CUL information providers, RAU coordinates and develops a knowledge base with processes, resources, and accessible measures. It drafts policies and procedures for approval by LMT and other bodies as appropriate. It produces/publishes results to advance evidence-based librarianship.

http://www.library.umass.edu/assessment/commchargeandmember.pdf

## Library Assessment Committee

**Charge:**
Advise and assist the Assessment Librarian with assessment initiatives of all types.

- Assist in the development of new assessment endeavors.
- Review, analyze, and discuss data for projects, such as LibQUAL+, or in response to specific questions.
- Support reaccredidation process as needed.

**Membership:**
Appointed by the Director of Libraries.

- Assessment Librarian.
- Up to four volunteers and/or nominees from all ranks of Library staff (at least one each from User Services and Collection Services).
- One Student Employee (if available).

**Chair:** Assessment Librarian.

**Term of Membership:** Two-year staggered terms, with staff rotating off each year on February 2. Members may be reappointed.

**Meetings:** Monthly or as needed, at the call of the Chair. Other members of the staff may be asked to join meetings when their specialized knowledge is needed.

**Information Sharing:** Maintain brief minutes of each meeting and publish them on LibWire. Provide additional information to the Director of Libraries as requested.

## Assessment Committee Members

Mike Allard
Rachel Lewellen, Chair
Stephen McGinty
Barbara Morgan
Emily Silverman

**I. Team Purpose**

Improve library decision making through oversight of processes involved in gathering, managing, and reporting, and analyzing a holistic data set for USC Libraries. Assessment areas encompass all library activities according to four data types:

1. User Perceptions: assesses customer satisfaction, and the impact of the library university research and learning. User perception data is collected through surveys, focus groups, and informal interviews.

2. Usage: assesses the use of library resources such as circulating materials, online catalogs, databases, and web pages; and services such as reference, instruction, and library spaces. Usage data is collected through a combination of system reports and self reporting.

3. Performance: assesses individual, departmental, program, team, and center output; process efficiency and quality control; project and goal completion; etc. Performance data is collected through a combination of system reports, self reporting, and manager reporting.

4. Descriptive: assesses the size of collections, number of personnel, amounts of budgets and expenditures, number and scope of services, etc. Descriptive data is collected through a combination of system reports and self reporting.

**II. Products & Services**

| *Services* | *Products* |
|---|---|
| Data Collection | Process maps, clearly defined data points, inputting instruction tutorials, online forms, cross-functional coordination |
| Data Management | Organization-wide management information system (Library Data Repository) |
| Data Reporting | Quarterly and annual reports, research survey participation, customized reports, presentations to groups |
| Assessment Training | Training for data gathering, data interpretation, decision making, turning data into action |
| Strategic Planning Support | External marketing research, USC needs assessments |

**III. Team Clients**

- Library Leadership
- Library Team Leaders, Coordinators, Center Chairs, and Department Managers
- USC Provost's Office
- USC School Deans, Academic Department Chairs
- USC Individual Researchers
- Non-USC Individual Researchers
- Non-USC Institutional Researchers (ARL, ACRL, etc.)

**IV. Key Results**

- Enhanced understanding of customer needs
- Informed managers, improved decision making
- Process quality control, enhanced efficiency, continuous improvement

**V. Guiding Principles**

- Data collected should provide feedback on the Library's mission and values
- Better data will improve performance

**VI. Time Frames**

- Monthly meetings
- Monthly data gathering
- Quarterly and annual reports
- Annual process review

**VII. Red Flag Issues**

- Data is not used (or not taken seriously)
- Data is overused (or taken too seriously)
- Data counted inaccurately
- Data not reported consistently
- Data not shared across the organization
- Data not stored centrally
- Managers untrained to interpret data
- Processes not documented
- Definitions not clear
- Collecting, reporting, and analysis roles not clear
- Self reporting where automation is possible
- Paper-based submission where electronic filing is possible

**VIII. Team Membership**

| | |
|---|---|
| Access Services | Brian Harlan |
| Administration Services | Game Quan |
| Collection Development | Lynn Sipe |
| Collection Management | Masoud Farajpour |
| Information Development Management | Wayne Shoaf |
| Integrated Document Delivery | Gabe Vincent |
| Instruction Services | Melanie Remy |
| Reference Services | Judy Truelson |

## UNIVERSITY OF TEXAS AT AUSTIN

http://www.lib.utexas.edu/vprovost/assessment/index.html

What Starts Here Changes The World
THE UNIVERSITY OF TEXAS AT AUSTIN

Libraries Home | My Account | Sitemap | Help

University of Texas Libraries

SEARCH: Library Web Site GO

About the Libraries | Research Tools | Library Services | Resources for You | Ask a Librarian

Home > About the Libraries > Library Administration > **Service Quality Assessment**

**Service Quality Assessment**
-About
-Assessment Plan
-LibQUAL+
-Library Summits
-Improvement Projects
-Quality Council

**Other Sites of Interest**
-Library Administration

### Service Quality Assessment and Improvement

We want to build a climate of continuous improvement in all areas of library service. All staff - reference librarians, web designers, book shelvers, and catalogers - should understand that they are part of an organic whole, a system, and that it is not enough to be concerned with the success of one's own department. Staff of a successful organization are concerned with the success of the entire organization.

*Fred Heath, Vice Provost*
*The University of Texas Libraries*
*The University of Texas at Austin*

**Contact Information**
Damon Jaggars, Associate Director, **jaggars@austin.utexas.edu**

Accessibility | Privacy/Confidentiality | Material Usage Statement
Comments | About This Site | Emergency Preparedness, Safety and Security
Page viewed: November 21, 2007 | Page last modified: November 6, 2007

UTDIRECT requires EID

http://www.lib.virginia.edu/mis/about/index.html

UNIVERSITY of VIRGINIA LIBRARY Management Information Services

About MIS
Mission Statement
Services Provided
Staff Members
Balanced Scorecard
Surveys & Reports
Statistics
Web Usability Testing
Links

# About MIS

## Mission Statement

Management Information Services (MIS) has responsibility for coordinating the collection and reporting of management information throughout the Library, and for advising Library administrators and managers on the uses of management information.

### The Department has the following functions:

1. To identify areas, functions, and tasks within the library where performance and service could be enhanced with management information data.

2. To develop expertise in various techniques of data collection and analysis, including focus groups and customer surveys, as well as analysis of computerized data files.

3. To educate the library staff in general, and library management in particular, as to the value of collecting and using management information.

4. To coordinate the statistical reports produced by the various library units, and to see that these reports correspond to the needs of the Library Administration and the appropriate national organizations.

5. To evaluate requests for management information and statistical reports, and to work with staff so that statistical reports and data analysis will be appropriate for the situation.

6. To assist the University Libraries Committee with their analyses and evaluations of library services and collections.

7. To provide the Library Administration with data for decision making by conducting and analyzing surveys and focus groups, etc.

Management Information Services
University of Virginia Library
PO Box 400114
Charlottesville, VA 22904-4114
phone: 434.924.3240 • fax: 434.924.1431

MIS Home • UVa Library Home
Search Library Website • UVa Home
Maintained by: MIS Webmaster
Last Modified: Friday, November 09, 2007

http://www.wsulibs.wsu.edu/general/WG/AWG.html

Washington State University

World Class. Face to Face. CAMPUSES WSU HOME WSU SEARCH myWSU

THE LIBRARIES Griffin Catalog / Article Indexes/E-Journals / Your Record/Renewals

- Libraries HOME
- Search/Site Map
- Hours/Phone
- Contact Us
- Ask a Question
- Interlibrary Loans
- Subject Resources
- Reference Shelf
- Classes & Handouts
- Assistive Technology
- About WSU Libraries
  - Services
  - Policies
  - Departments
- New Books
- Libraries FAQ
- News/Exhibits/Events
- Other Library Catalogs
- Giving to the Libraries
- For Library Employees

The Libraries - Your Information & Resources Gateway

## ASSESSMENT WORKING GROUP

**ROLE:**

- To address issues related to assessing library collections, technologies and services
- To initiate assessments throughout the Libraries, chiefly based on client input
- To coordinate with and among working groups conducting assessments according to their charges
- To recommend to Library Council (LC) and/or other working groups actions based on results of assessments

The Assessment Working Group is intended to be an oversight body, with subgroups (task forces) on specific issues in assessment which include WG members and nonmembers who are interested in a particular aspect of assessment. The Group will also coordinate efforts of other working groups and act as a clearinghouse. Branch campuses may be included to the extent they wish.

**MEMBERSHIP:**

- Assistant Director for Public Services and Outreach (ex officio)
- Library Webmaster (ex officio)
- Library Data Officer (ex officio)
- 6 Library faculty or staff, including at least 2 reference librarians, chosen by LC after soliciting volunteers

Terms will be for two years, starting in July. (Initial appointments included 3 people with 2-year terms and 3 people with 3-year terms to stagger turnover.)

**GOVERNANCE:**

The Assessment Working Group reports to the Director. The working group will select its own chair for a minimum six month term of office and will meet regularly or as needed. The group will coordinate its activities with other working groups and library departments, as warranted. Working groups will establish procedures and make policy within the scope of their responsibilities. Working groups are accountable for their decisions and are expected to share decisions with library colleagues using the standard working groups procedure. Any proposals generated by the working group that require non-materials budget funding for implementation must be referred to Library Council for funding approval. Similarly, any proposals with very broad policy implications also need Library Council approval. The chair of the working group is expected to update Library Council on its activities on a regular basis.

**SPECIFIC CHARGE:**

- To create an overall assessment plan for the WSU Libraries
- To undertake and coordinate assessments, surveys, etc. aimed at measuring the quality of our services and collections, and the needs and expectations of our clientele
- To ensure assessments are undertaken in accordance with University regulations
- To report the results of said assessments to LC and other working groups
- To make recommendations for changes based on the results of said assessments
- To coordinate WSU's participation in national and regional library assessments of services and collections. This does not include quantitative statistical reports.
- To provide a repository for the results of assessments done by WSU Libraries groups
- To initiate information and training sessions related to assessment
- To disseminate information on assessment initiatives

**DEFINITIONS:**

- Assess – to determine the importance, size, or value of something
- **Coordinate** – to bring into common action, movement, or condition: harmonize

Approved by LC: 11-8-00; revised membership 11-15-00; revised 9-11-02; revised membership 06-12-06

Decisions Digest
Minutes Archive
List Serve Information
Working Agreement
Assessment Web Site
Strategic Plan
Five Year Assessment Plan (2002)

Contact us: libweb@wsu.edu 509-335-9671 | Accessibility | Copyright | Policies
WSU Libraries, Washington State University, Pullman, WA, 99164-5610 USA

*Last updated on Monday June 12th, 2006.*
http://www.wsulibs.wsu.edu/general/WG/AWG.html

# Organization Charts

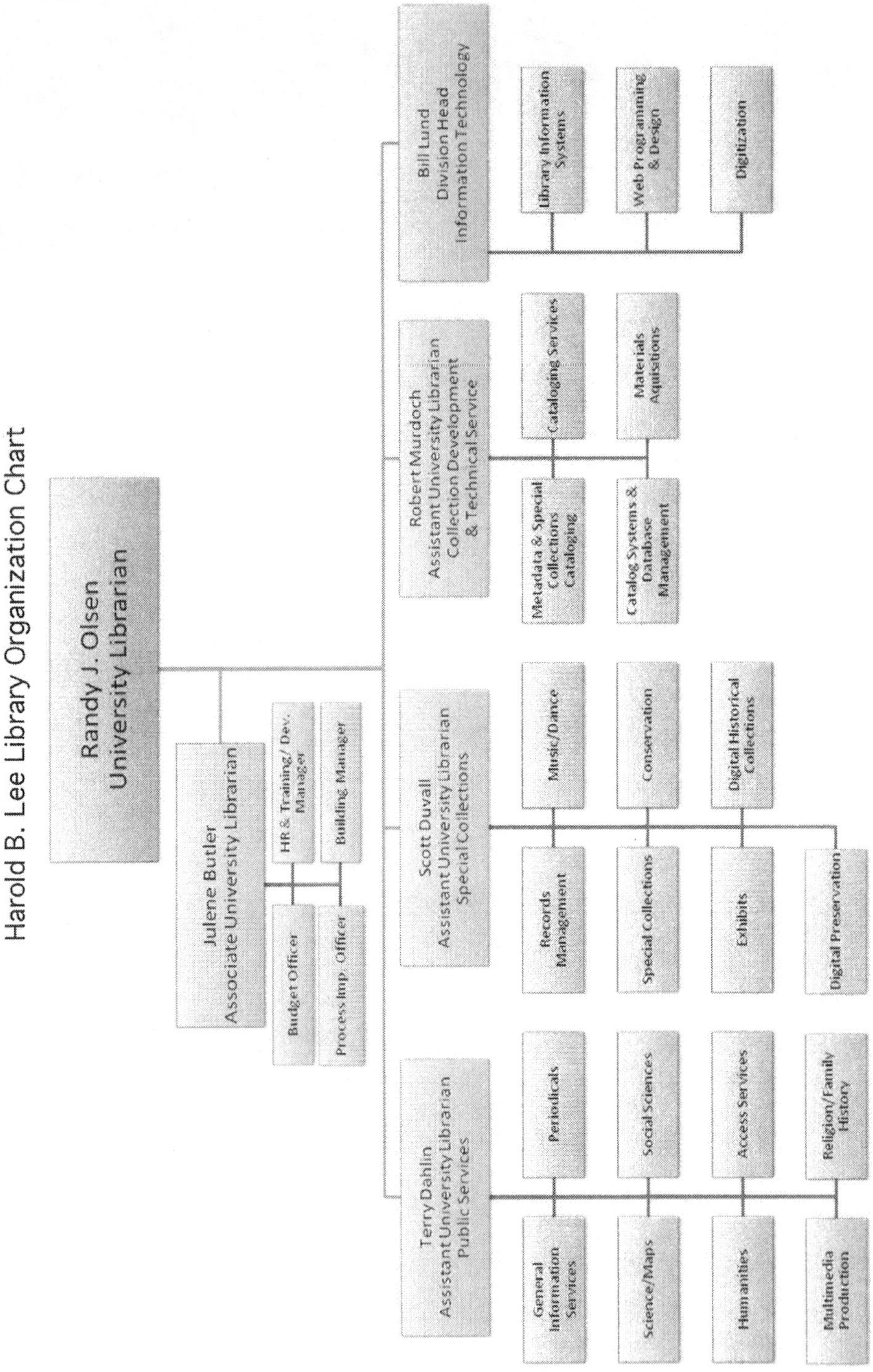
Harold B. Lee Library Organization Chart
Randy J. Olsen
University Librarian
Julene Butler
Associate University Librarian
HR & Training/ Dev. Manager
Building Manager
Budget Officer
Process Imp. Officer
Terry Dahlin
Assistant University Librarian
Public Services
General Information Services
Periodicals
Science/Maps
Social Sciences
Humanities
Access Services
Multimedia Production
Religion/Family History
Scott Duvall
Assistant University Librarian
Special Collections
Records Management
Music/Dance
Special Collections
Conservation
Exhibits
Digital Historical Collections
Digital Preservation
Robert Murdoch
Assistant University Librarian
Collection Development
& Technical Service
Metadata & Special Collections Cataloging
Cataloging Services
Catalog Systems & Database Management
Materials Aquisitions
Bill Lund
Division Head
Information Technology
Library Information Systems
Web Programming & Design
Digitization

UCIrvine | LIBRARIES

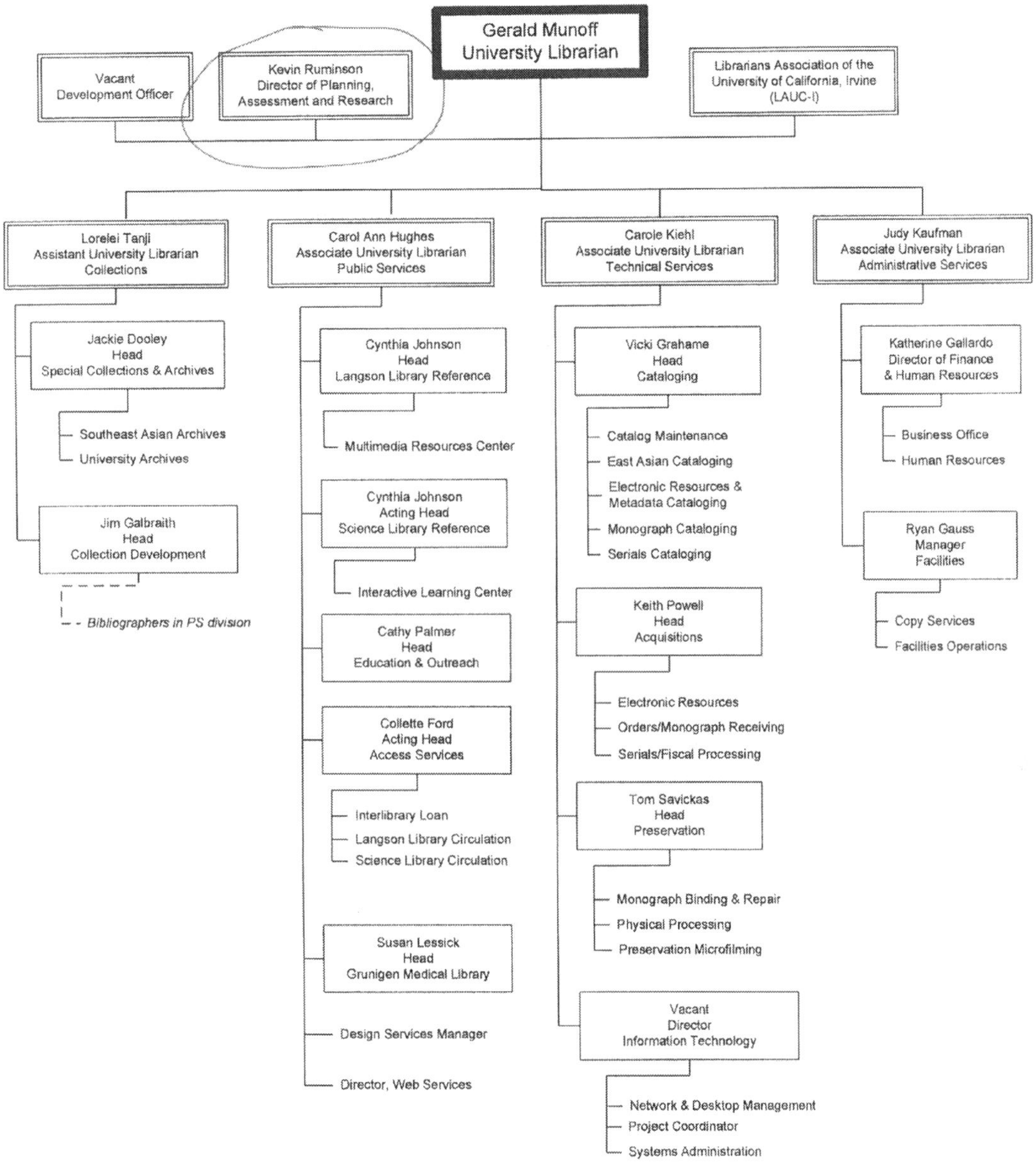


May 8, 2007

http://www.lib.uchicago.edu/e/about/orgchart.pdf

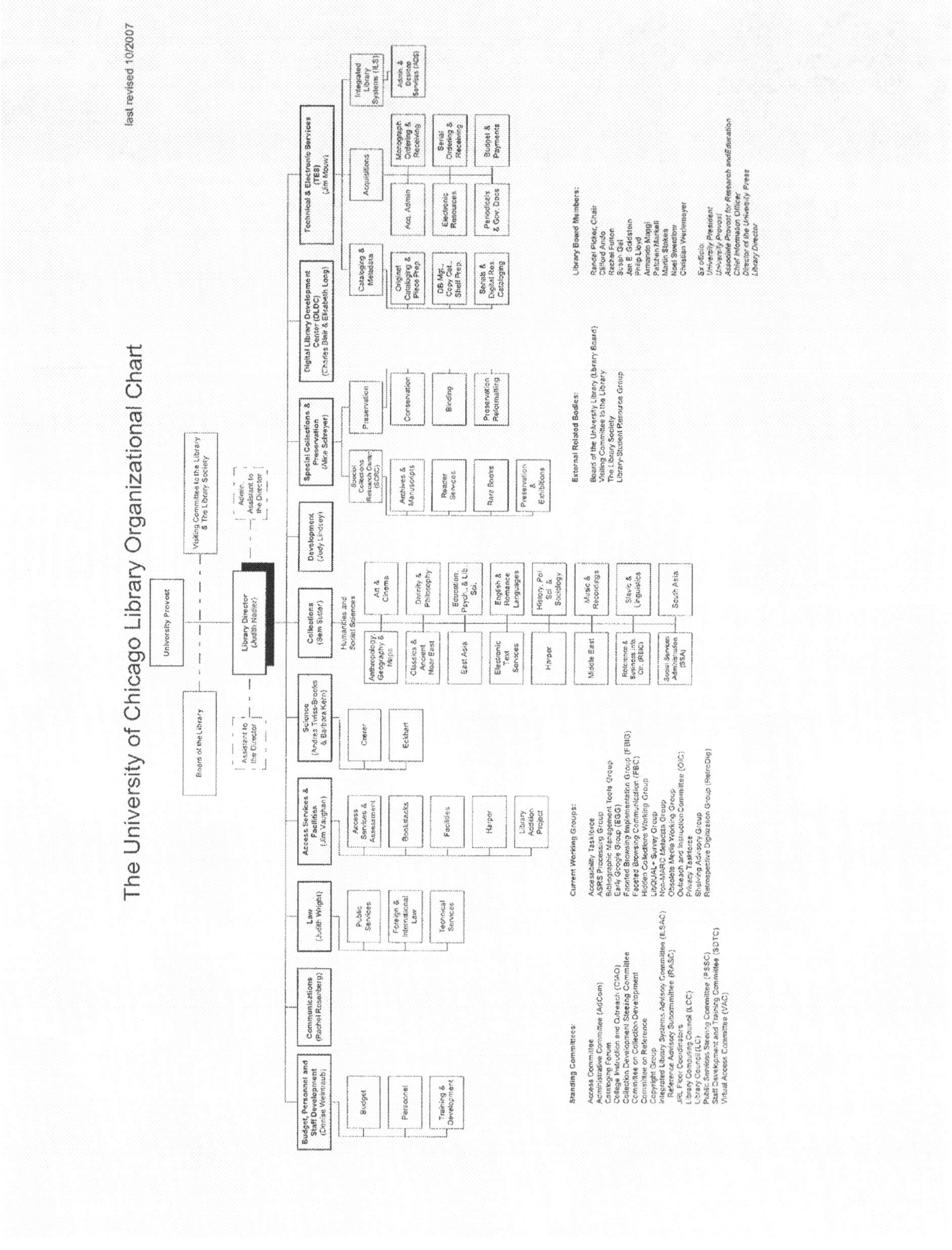

http://www.library.cornell.edu/iris/documents/PSAOrgChart.pdf

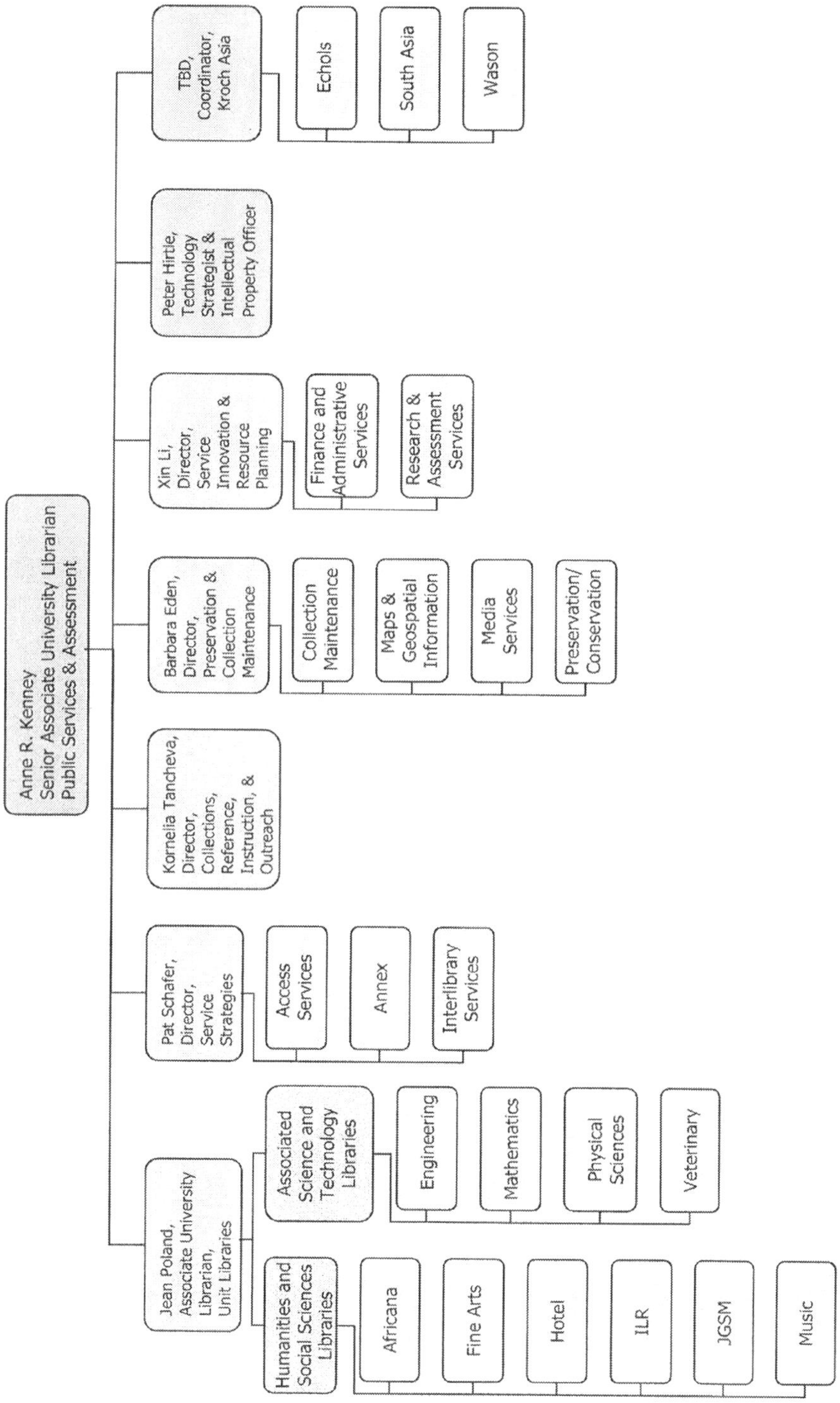

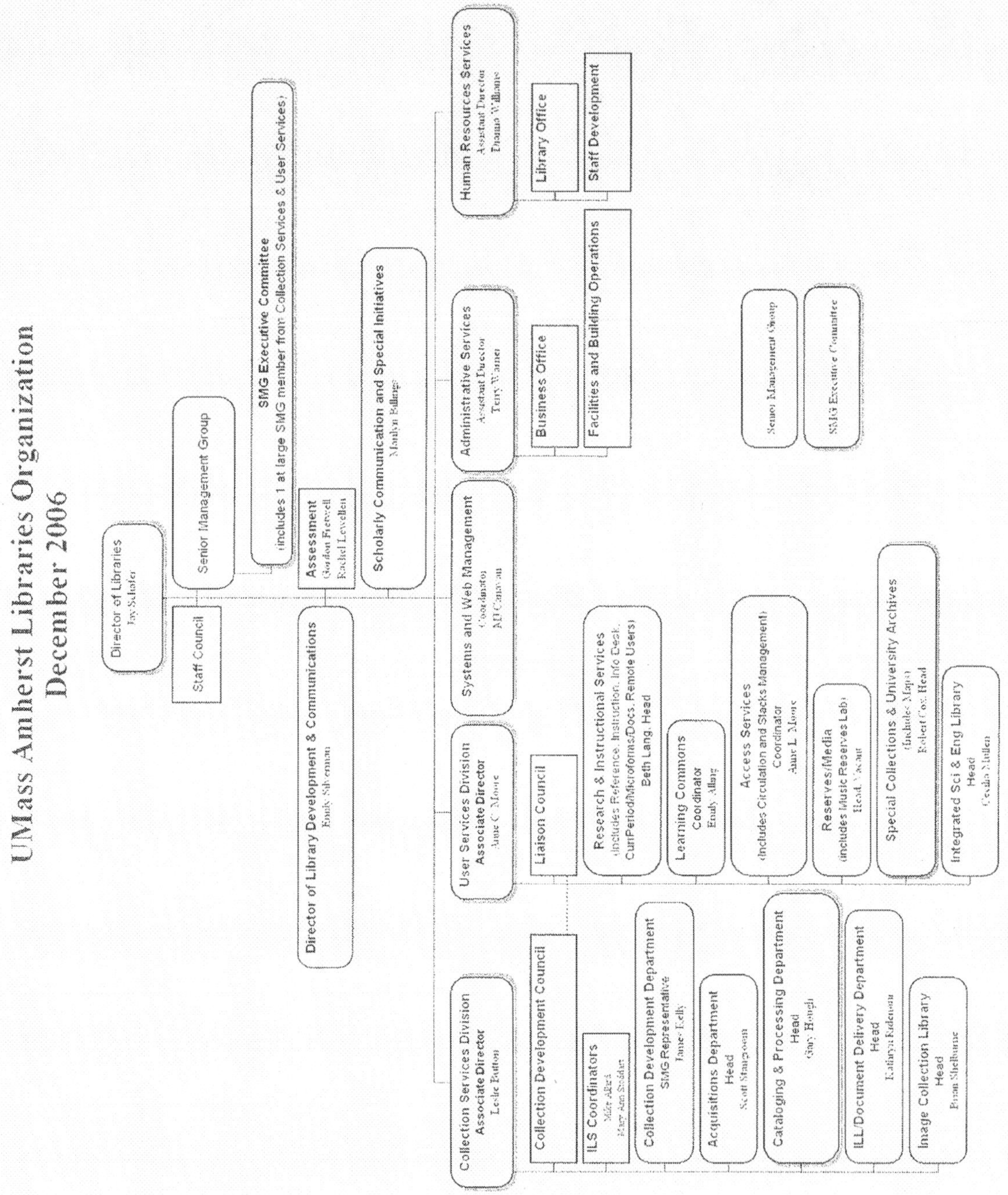
UMass Amherst Libraries Organization
December 2006
Director of Libraries
Staff Council
Senior Management Group
SMG Executive Committee
(includes 1 at large SMG member from Collection Services & User Services)
Assessment
Scholarly Communication and Special Initiatives
Director of Library Development & Communications
Collection Services Division
Associate Director
Leslie Button
User Services Division
Associate Director
Systems and Web Management
Coordinator
Administrative Services
Assistant Director
Human Resources Services
Assistant Director
Collection Development Council
ILS Coordinators
Collection Development Department
SMG Representative
Acquisitions Department
Head
Cataloging & Processing Department
Head
ILL/Document Delivery Department
Head
Image Collection Library
Head
Liaison Council
Research & Instructional Services
(includes Reference, Instruction, Info Desk, CurrPeriod/Microforms/Docs, Remote Users)
Beth Lang, Head
Learning Commons
Coordinator
Access Services
(includes Circulation and Stacks Management)
Coordinator
Reserves/Media
(includes Music Reserves Lab)
Special Collections & University Archives
Integrated Sci & Eng Library
Head
Business Office
Facilities and Building Operations
Library Office
Staff Development
Senior Management Group
SMG Executive Committee

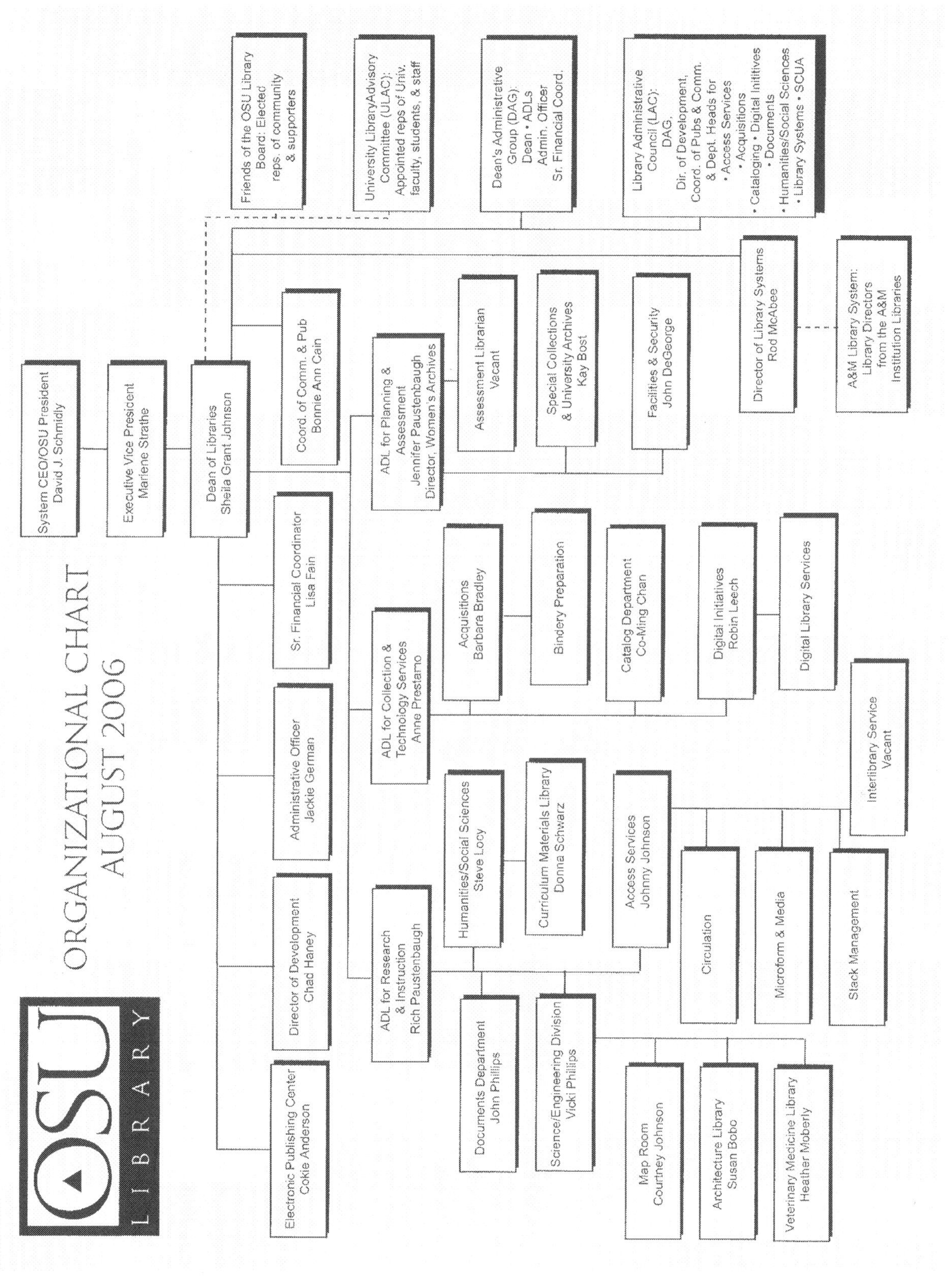
OSU LIBRARY
ORGANIZATIONAL CHART
AUGUST 2006
System CEO/OSU President David J. Schmidly
Executive Vice President Marlene Strathe
Dean of Libraries Sheila Grant Johnson
Friends of the OSU Library Board: Elected reps. of community & supporters
University LibraryAdvisory Committee (ULAC): Appointed reps of Univ. faculty, students, & staff
Dean's Administrative Group (DAG): Dean • ADLs Admin. Officer Sr. Financial Coord.
Library Administrative Council (LAC): DAG, Dir. of Development, Coord. of Pubs & Comm. & Dept. Heads for • Access Services • Acquisitions • Cataloging • Digital Inititives • Documents • Humanities/Social Sciences • Library Systems • SCUA
Director of Library Systems Rod McAbee
A&M Library System: Library Directors from the A&M Institution Libraries
Coord. of Comm. & Pub Bonnie Ann Cain
ADL for Planning & Assessment Jennifer Paustenbaugh Director, Women's Archives
Assessment Librarian Vacant
Special Collections & University Archives Kay Bost
Facilities & Security John DeGeorge
Sr. Financial Coordinator Lisa Fain
ADL for Collection & Technology Services Anne Prestamo
Acquisitions Barbara Bradley
Bindery Preparation
Catalog Department Co-Ming Chan
Digital Initiatives Robin Leech
Digital Library Services
Administrative Officer Jackie German
Director of Development Chad Haney
Electronic Publishing Center Cokie Anderson
ADL for Research & Instruction Rich Paustenbaugh
Humanities/Social Sciences Steve Locy
Curriculum Materials Library Donna Schwarz
Access Services Johnny Johnson
Interlibrary Service Vacant
Circulation
Microform & Media
Stack Management
Documents Department John Phillips
Science/Engineering Division Vicki Phillips
Map Room Courtney Johnson
Architecture Library Susan Bobo
Veterinary Medicine Library Heather Moberly

http://www.lib.utexas.edu/vprovost/lib_info/images/20070212_viceprovost.pdf

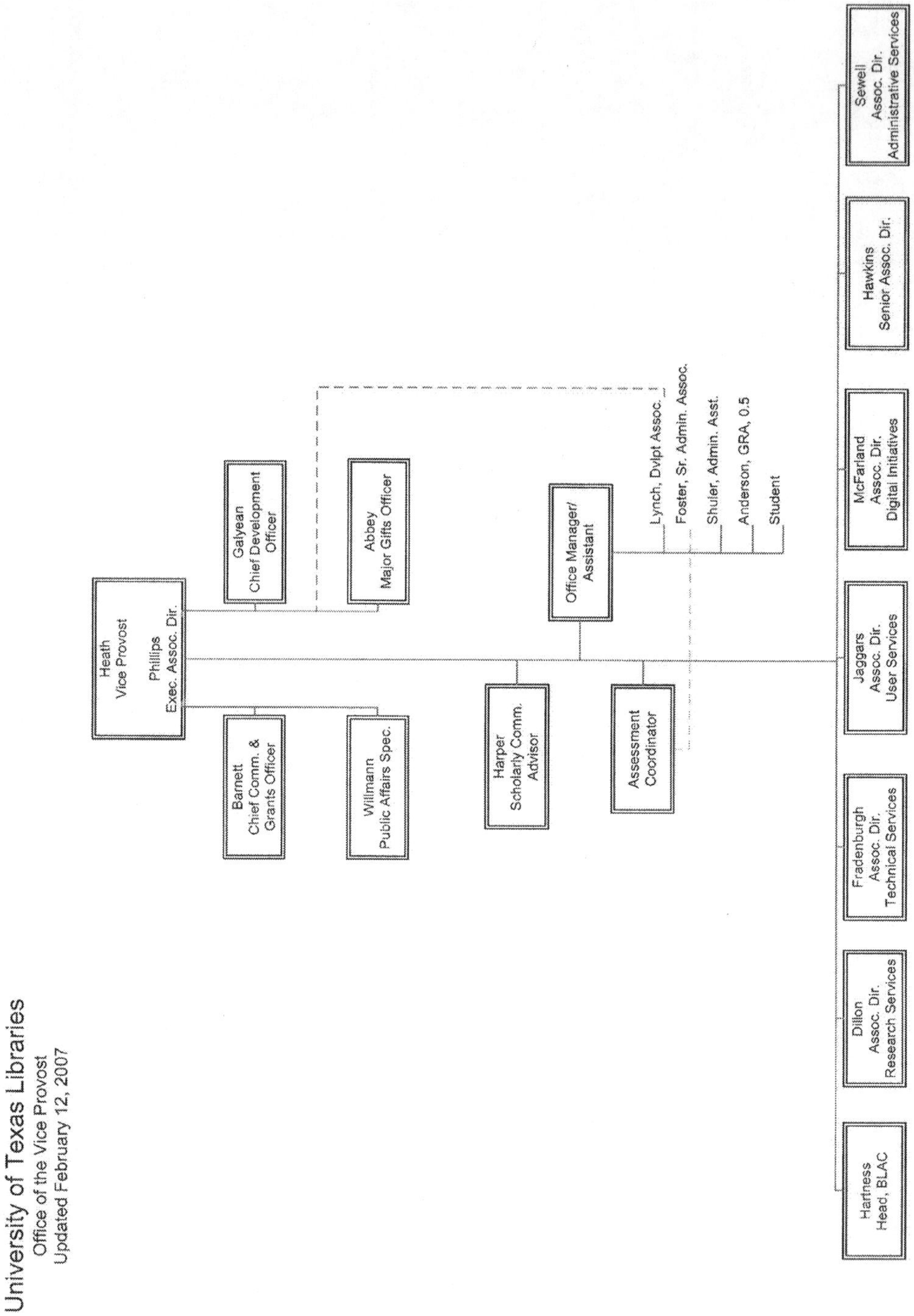

# Assessment Web Sites

http://ucblibraries.colorado.edu/internal/assessment/index.htm

University of Colorado at Boulder | CU: Home | Search | A to Z | Map

UniversityLibraries

Contact Us
Search this Site
Libraries Home

Research Resources | About the Libraries | Services  Chinook Catalog  Keyword  Search

**Assessment Committee**

- Home
- Membership
- LibQUAL+™ at CU
- Responsibilities
- Data Sets
- Assessment Projects
- External Resources
- Bibliography
- Personnel
- Assessment Websites

About the Libraries > Libraries & Departments >

## Assessment Committee

### Charge

To foster a culture of evidence-based practice, the Libraries Assessment Committee will promote and initiate ways by which measurement and evaluation can be integrated into existing library management practices.

The Committee will initiate and/or coordinate the regular measurement of such subjects as patron use of library services, quality of collections, use of technology, and impacts of technical operations. The Committee will undertake and/or coordinate measurements of the quality of services and collections and the needs and expectations of the Libraries' clientele to support strategic planning, facilities planning, budget or performance review cycles, vendor contract renewals, or technology/systems conversions. It will endeavor to conduct library assessment in relation to institutional goals and outcomes. The Committee will recommend to Library Council and/or other committees actions based on the results of measurement and assessment.

Colorado
University of Colorado at Boulder

University Libraries, 184 UCB, 1720 Pleasant Street, University of Colorado, Boulder, CO 80309-0184
Libraries Information 303-492-8705 | Reference & Instructional Services 303-492-7521
Send Comments to the Libraries
© 2007, The Regents of the University of Colorado

http://www.library.uiuc.edu/assessment/index.html

I UNIVERSITY LIBRARY

University of Illinois at Urbana-Champaign » University Library » Library Assessment @ UIUC

# ASSESSMENT @ UIUC LIBRARY

Welcome to the University of Illinois Library Assessment Website. Assessment is helping our library to be more responsive. The Services Advisory Committe (http://www.library.uiuc.edu/committee/charges/advise.htm) is overseeing assessment activities and welcomes your suggestions.

## FACTS @ A GLANCE (facts.html)

UIUC Library unit annual reports by fiscal year, University of Illinois quick facts, and Library Assessment Working Group (LAWG) minutes (facts.html)

## ARL COMPARISON STATISTICS (arl.html)

ARL resources including statistical annual reports, publications and comparative graphs and tables (arl.html)

## LIBRARY SURVEYS (surveys.html)

Library surveys including introductory documents, recent student and faculty surveys, LibQUAL+ and user surveys (surveys.html)

## DEPARTMENTAL LIBRARIES (departmental.html)

Check out statistical information being collected by departmental libraries (departmental.html)

## WEB RESOURCES & TOOLS (tools.html)

Online tools to assist in the research process of compiling statistics and forming surveys, includes survey builder tools (tools.html)

## BIBLIOGRAPHY (biblio.html)

Readings on assessment, books on survey research methods and related Library faculty publications (biblio.html)

## USEFUL LINKS (links.html)

Online resources including the National Center for Education Statistics and peer institution library assessment (links.html)

## FEEDBACK (feedback.html)

Please send us your comments and suggestions for this site (feedback.html)

This site was developed by Merinda Hensley, Lisa Hinchliffe, and Bob Burger with resources and feedback from throughout the University Library. Funding was provided by the University Library Executive Committee.

http://library.osu.edu/sites/staff/assess/

The Ohio State University www.osu.edu Help Campus map Find people Webmail Search Go

University Libraries [subheading ]

OSUL Home | Find | Borrow | About OSUL | Libraries | Learn | Off-campus Sign-in | My Record | Help

Members

Andersen, Lila, 6/30/2009
Burns, Beth, 6/30/2009
Cetwinski, Tom Co-Chair, 6/30/2010
Feather, Celeste, *ex officio*
Hall, Kathy, *ex officio*
Helmick, Nancy Co-Chair, 6/30/2010
Kissel, Laura, 6/30/2008
Murphy, Sarah, 6/30/2010
Opperman, Bill, 6/30/2008
Pearson, Penny, *ex officio*
Powell, Carol, 6/30/2008
Rahnema, Sue, 6/30/2010

## Assessment Committee

The charge to the Committee includes, but is not limited to:

- Creating an atmosphere and culture of assessment in the Libraries
- Analyzing statistical data that is gathered on a continuing basis in various departments/offices
- Conducting systematic and ongoing assessments of the Libraries' services, facilities, and operations using a variety of survey and assessment tools.

Information gathered will be used to identify trends in collections, services, and use within University Libraries and in comparison with our peer institutions and other ARL libraries and develop recommendations for meeting the Libraries? Service Values.

### Reports from Surveys/Questionnaires

- 2007 LibQUAL
- 2006 Exit Survey
- 2006 LibQUAL
- 2004 OSU Poll
- 2001 Exit Survey
  - Overview
  - Main Library

### Annual Committee Reports

- April 10, 2007
- April 11, 2006
- April 12, 2005
- April 7, 2004

### Presentations

- ServQUAL (ASC Meeting, March 13, 2006)
- LibQual (ASC meeting, March 13, 2006)
- Assessment Committee Activities (General Staff Meeting, March 6/7, 2006)
- Assessment in the Libraries (Regional Campus Meeting, February 23, 2006)

### Minutes of Meetings

- September 25, 2007
- May 22, 2007
- April 24, 2007
- February 27, 2007
- January 23, 2007
- December 15, 2006
- October 27, 2006
- September 22, 2006
- August 25, 2006
- August 11, 2006
- July 14, 2006
- June 9, 2006
- May 26, 2006
- May 12, 2006
- April 14, 2006
- February 24, 2006
- January 27, 2006
- December 16, 2005
- December 2, 2005
- November 18, 2005
- October 28, 2005
- September 9, 2005
- August 26, 2005
- August 5, 2005

Send comments or questions to Betty Sawyers
Last modified: 6/28/2007


1858 Neil Avenue Mall

http://metrics.library.upenn.edu/prototype/datafarm/

About the Data Farm | Shared File Depository | Vendor Statistics | Related Sites | Staff Web

# Data Farm

University *of* Pennsylvania Library

Penn Library Facts 2006

## Tools and Data

E-resource Tracking (use measures for e-journals, databases, and other electronic resources)

Selected Counter Data (the interface loads slowly)
Vendors and coverage dates

Gate Counts

Photocopier & Printer Use

Image Collection Use (down pending log changes)

BorrowDirect Data Repository

EZBorrow Data Repository

Public Web Log Analytics (down pending log changes)

Research & Instructional Services

Annual Data Collection

2004 Library Quality/Impact Survey

2002 Library Quality/Impact Survey

Graduate Student Workshops

Annual Survey of Who Asks Questions Where

Library Staff Census (under construction)

## Voyager Reports

Voyager Funds - Report Builder

Circulation | Collection Management Reports

Report: Items with no circulation since the introduction of Voyager. Available for specific locations and classifications

Report: List call numbers and holdings records (mfhd) without items in Voyager. Available for Van Pelt, Lippincott and East Asia

Report: Collection Inventories. Item by item report for perm_location, includes http://metrics.library.upenn.edu/prototype/datafarm/

bibliographic details, detailed circ history and item status.

Bib-Action | MFHD-Action Reports [contact M. Rohdy for info.]

---

**Test Folder**

Webdoc Analysis

---

**Web Trends Reports** [tests and samples, not consistently updated]

Visits to www.library

Websites By Subject
(pre 2002 web redesign)

Historical Corporate Reports

Online Books Page

Link Analysis

Alumni Portal

BlackBoard Summary

http://www.lib.virginia.edu/mis/

UNIVERSITY of VIRGINIA LIBRARY **Management Information Services**

Management Information Services (MIS) has responsibility for coordinating the collection and reporting of management information throughout the Library, and for advising Library administrators and managers on the uses of management information.

About MIS
Balanced Scorecard
Surveys & Reports
Statistics
Web Usability Testing
Links

Management Information Services
University of Virginia Library
PO Box 400114
Charlottesville, VA 22904-4114
phone: 434.924.3240 • fax: 434.924.1431

UVa Library Home
Search the Library Site • UVa Home
Maintained by: MIS Webmaster
Last Modified: Friday, November 09, 2007

http://lib.washington.edu/assessment/

UNIVERSITY OF WASHINGTON Libraries

Libraries Home | Resources | Services | About

ask us email | chat | phone

off-campus access (log in)

Libraries Home > **Library Assessment Group**

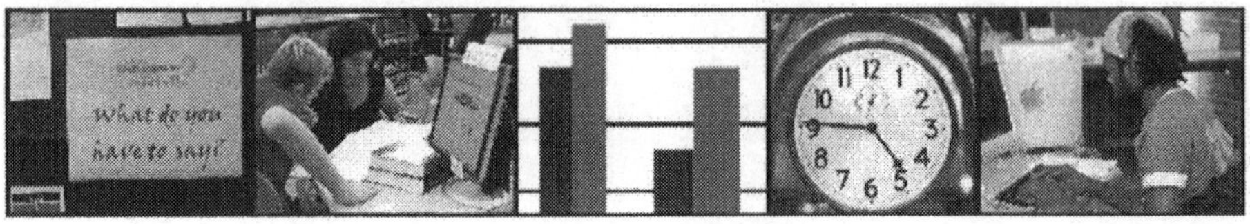


# UW Libraries Assessment

Libraries Assessment Group | UW Libraries Triennial Survey | Other Surveys | Activities

## Library Assessment Group

The Library Assessment Group works with Steve Hiller, Director of Assessment and Planning, to initate and support library assessment efforts within the University Libraries.

Members

## UW Libraries Triennial Survey

The University of Washington Libraries has conducted extensive large-scale surveys of faculty and students since 1992. These triennial surveys focus on library use and satisfaction as well as user needs and library priorities.

2007 Survey Information

Previous Triennial Surveys and Results

## Other Surveys, Results & and Assessment Info

In Library Use Surveys

LibQual

Usability Testing

## Library Statistics

Selected Library Statistics

User Query Sampling

Circulation Statistics

Monthly Gate Counts (.xls)

Gate Count Summary (.xls)

## Activities

Presentations

http://www.wsulibs.wsu.edu/Assessment/default.html

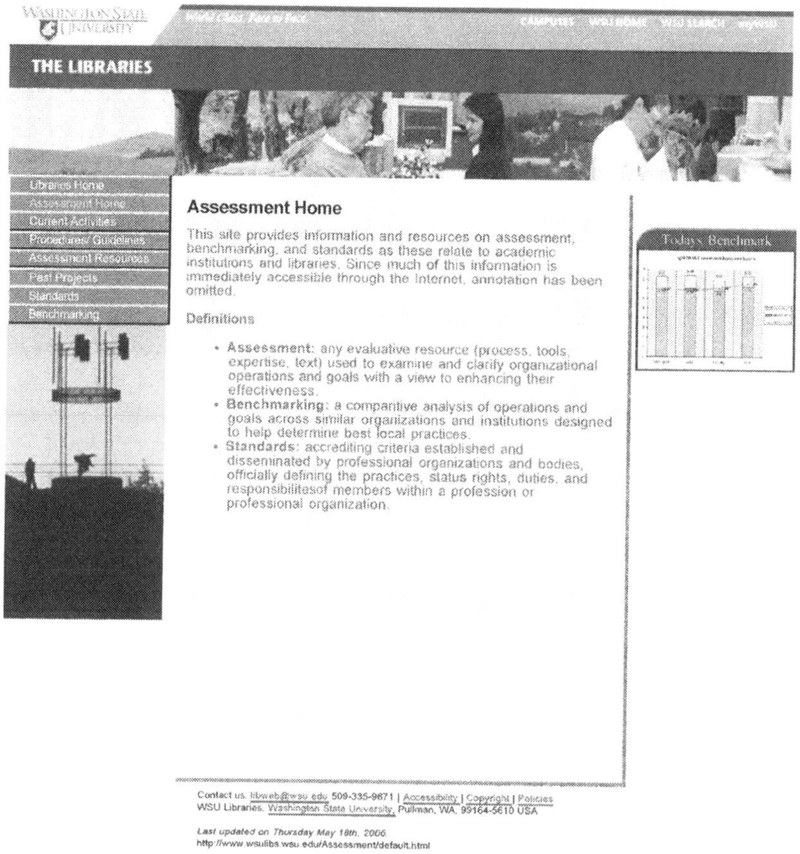
WASHINGTON STATE UNIVERSITY

THE LIBRARIES

Libraries Home
Assessment Home
Current Activities
Procedures/ Guidelines
Assessment Resources
Past Projects
Standards
Benchmarking

## Assessment Home

This site provides information and resources on assessment, benchmarking, and standards as these relate to academic institutions and libraries. Since much of this information is immediately accessible through the Internet, annotation has been omitted.

Definitions

- **Assessment:** any evaluative resource (process, tools, expertise, text) used to examine and clarify organizational operations and goals with a view to enhancing their effectiveness.
- **Benchmarking:** a comparitive analysis of operations and goals across similar organizations and institutions designed to help determine best local practices.
- **Standards:** accrediting criteria established and disseminated by professional organizations and bodies, officially defining the practices, status rights, duties, and responsibilitesof members within a profession or professional organization.

Todays Benchmark

Contact us: libweb@wsu.edu 509-335-9671 | Accessibility | Copyright | Policies
WSU Libraries, Washington State University, Pullman, WA, 99164-5610 USA

*Last updated on Thursday May 18th, 2006.*
http://www.wsulibs.wsu.edu/Assessment/default.html

Western Libraries Intranet

# Assessment Committee

## WESTERN LIBRARIES' ASSESSMENT PROJECTS - USERS

***In Progress***

Taylor Library EBL E-Book Survey

***Completed***

Service Quality Survey LibQUAL+™ 2007
MultiSearch Feedback Survey Useability Testing: Library Catalogue Display Information
Western Libraries Chat with Us Online Survey
Scholars Portal MultiSearch Usability Testing
Taylor Graduate Students' Library Research Skills Needs Assessment Survey (summer2006)
Information Discovery (Info Disco) Working Group Questionnaire
Web Site Steering Committee, 2005 Feedback Team
Project SAILS - Phase III
Scholars Portal Evaluation
Website Redesign Feedback (July 2004)
LibQUAL+™ March 2004
Western Libraries Website User Survey
*Ad Hoc* Working Group on Web Site Feedback

## In Progress

| | |
|---|---|
| **Project Name** | **EBL E-Book Survey** |
| **Team Names** | Taylor Library E-Book Working Group (Harriet Rykse, contact) |
| **Purpose** | To obtain user perceptions of e-book use in general and EBL e-books specifically |
| **Timing** | December 2006 until enough data is collected to draw some valid conclusions |
| **Sample Group** | Anyone who accesses an EBL e-book |
| **Sample Size** | n/a |
| **Methodology** | A Web survey setup with four questions for users of EBL e-books. The survey pops up the second time a user accesses an EBL e-book. The person trying to access the e-book has an option to answer immediately, answer later or never answer. Once he/she has answered, they will not see the surgvey again. After answering the questions, the person is linked to the e-book initially requested. |

# Assessment Activity Reports

**Information Question Logging Project**
**November 2006**

The Information Logging Project was conducted from September 11-24, 2006 throughout all public service points in the University of Alberta Libraries system: reference and circulation desks, roving reference, satellite reference, and reference staff offices. Questions were summarized and logged by staff, noting unit library and service point, question format (in-person, instant messaging, email, and phone), and approximate time taken to answer the question.

The key objective of the pilot was to:

- Help inform additions and edits to the web FAQ
- Help identify policies and procedures that users find problematic

Report Recommendations Summary:

1. Identified Policy and procedure issues be addressed by relevant Library committees (Appendix A)
2. LOG should address issue of whether there should be consistency in equipment available in all libraries (e.g. colour printer, scanner, headphones)
3. Individual libraries should review spreadsheet of logged questions for library specific issues

DATA COLLECTION

The timing of the reference question logging was planned for the first two weeks of classes, assuming that questions during this time period would tend to focus on the information needs of students and faculty during the first week of classes (for example, directional questions that could be addressed with better signage, or questions that could be incorporated into the online library FAQ).

Over the period, a total of 7,014 questions were logged from the various service points. It should be noted that one circulation service point mistakenly began the question logging a week into the data collection period; however, the questions received by this service point totaled just over 1% of all questions received, which would not significantly affect the final dataset.

The Assessment Librarian categorized all questions using a slight variation of the KKAR Classification Scheme, utilized by Rutgers (See Appendix B) to code and organize the questions for analysis. Categories used were Directional, Ready Reference, Research, Policy and Procedural, Inappropriate, No Question, and Holdings; the Subject Search category was collapsed into the Research category.

DATA ANALYSIS

This analysis has a system-wide focus, looking at results that are generalizable across the University of Alberta Libraries. The categorized coded spreadsheet of questions has been distributed to individual unit libraries to enable them to address local issues.

Where/How Were Questions Asked?

The majority of the questions received were received at reference and circulation desks, as seen in Fig. 1 below.

Fig. 1: Questions by Service Point (Percentage of All Questions; All Libraries; n=7,014)

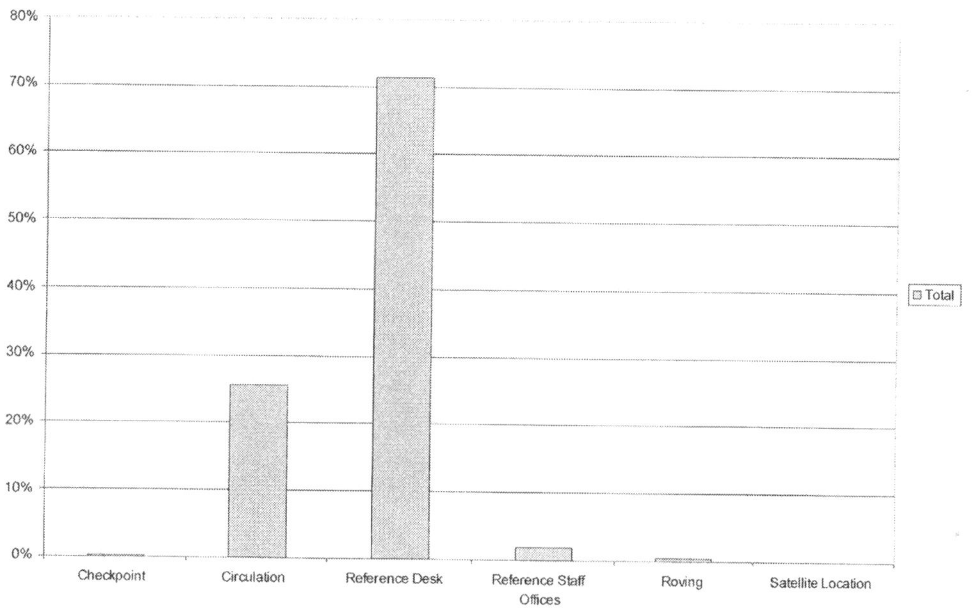


Further, questions during the collection period were most likely to be asked in-person than through phone, email, or instant messaging formats. Fig. 2 shows the question format for the two key service points, reference and circulation desks. Circulation desks did not receive questions in either email or instant messaging format at all during this time; it is assumed that later in the term, circulation desks would have received many more questions. Only reference desks received questions via email or instant messaging. Interestingly, in DeGroote's study of health science library reference questions during the academic term (collected during the month of November 2003), the majority of questions also tended to be asked in person (2005).

8

Fig. 2: Questions Asked by Format at Circulation and Reference Desks
(Percentage of Total Questions; n=6,824)

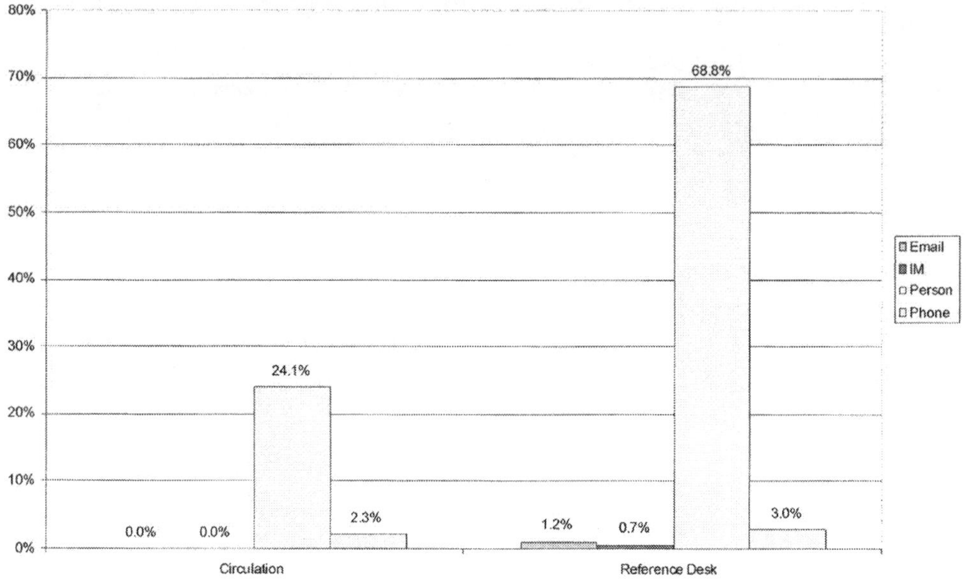


The majority of questions asked at all service points were in the Policies and Procedures category (30.4%), followed by Directional questions (20.5%), Technology questions (17.1%), and Holdings (16.7%). Reference and Ready Reference questions were 13.7% and 1.7% of the total questions asked, respectively. De Groote notes that past studies of reference questions have resulted in directional questions composing between 30-35% of all questions asked at the reference desk (2005). Again, timing of data collection is likely a factor here.

9

Fig. 3A: Categories of Questions Asked, All Service Points (n=6,974)

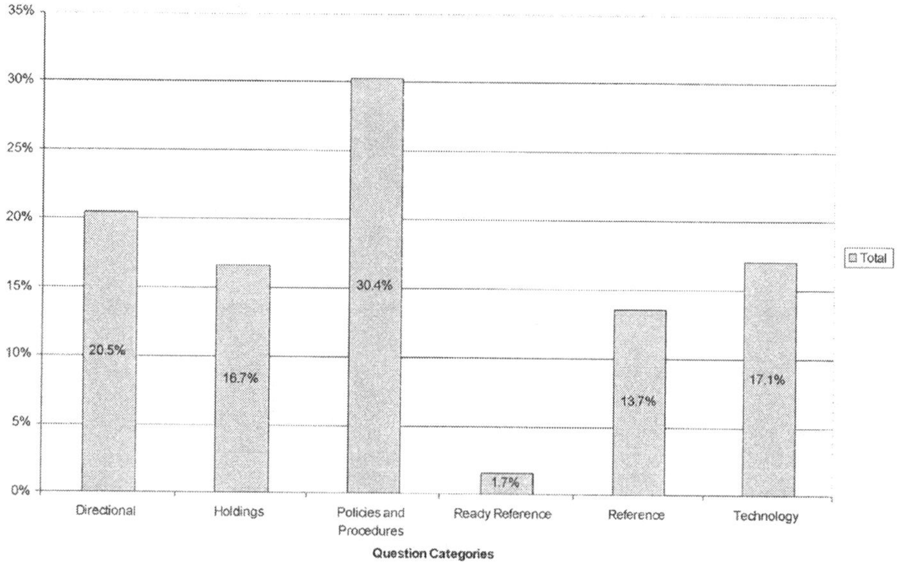


Looking more closely at the types of questions received by these two main types of service points, we see that reference desks received more questions in every category (fig. not shown). Questions categorized as Policies / Procedures and Directional were most frequently asked at both types of service points.

Policy/Procedure Questions

Users questioned and /or sought clarification on a number of policies and procedures. Most policy and procedure issues were related to:

Circulation policies/rules
Request service
Fines
Library privileges
Onecard
Availability of equipment/technology
Facilities

Appendix A identifies these issues. It is recommended that these issues be reviewed by the relevant Library committee. In addition, in response to numerous information requests on copyright, a copyright FAQ is currently being developed with Cindy Paul.

10

There is considerable variability of equipment available amongst the libraries. (e.g. colour printer, scanner, headphones, DVD burner). This is an issue that should be addressed by LOG.

Directional Questions:

A more specific categorization and analysis was undertaken for directional questions. Directional subcategories included:

- Call Number
- Circulation Desk
- Computer Labs (including questions about other PAC stations)
- Library Directions (including other campus libraries, directions to specific floors, study space, and directions to course sign-up sheets, staff, book return, supplies / equipment, etc.)
- Other (change for copying, where to make a phone call, etc.)
- Outside Library (non-library destinations on campus or in the city)
- Printer / Photocopier / Onecard Machine
- Room (specific numbered rooms)

The Reference Services Team's Working Group on Web Content has reviewed all of the directional questions and added information and Library maps to the Libraries FAQ.

The above sub-categorization demonstrates a very specific focus, with one result being that proportionally few questions were asked in most of the categories. Fig. 5 shows the complete count of these sub-questions for all library units.

**Fig. 5: Type of Directional Questions, All Libraries (n=1,381)**

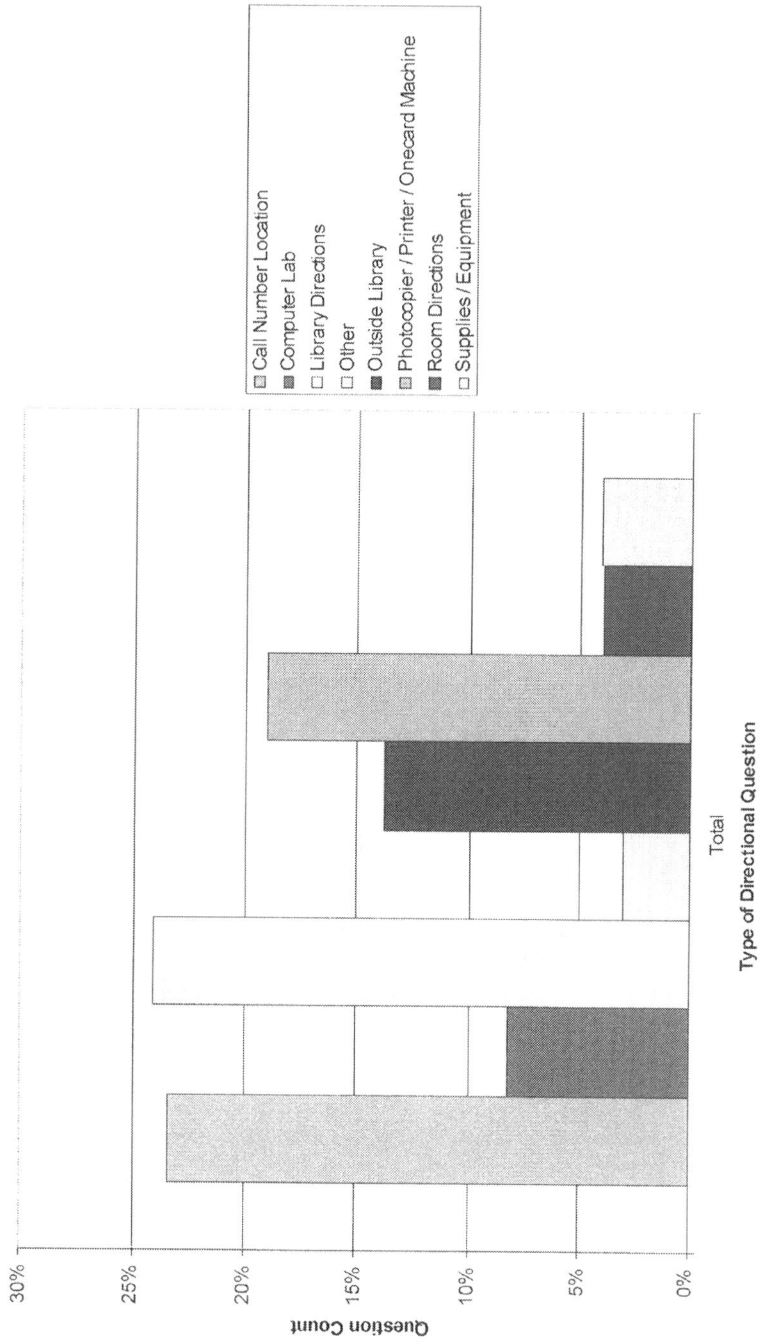


12

The top 5 types of directional questions were analyzed by unit libraries (both large units and medium / small).

Figures 6A and 6B, and 7A and 7B look at total directional questions asked at the large unit libraries and the medium / small unit libraries, respectively. Directional questions were most likely to be asked at the Education and Sci/Tech libraries. The Health Sciences library was least likely of all large units to be asked these questions; therefore, other unit libraries may look at the Health Sciences library signage to see specifics of placement, contrast, and other visibility factors for the different types of directional questions asked.

Fig. 6A: Total Directional Questions by Large Unit Library
(n=4,973)

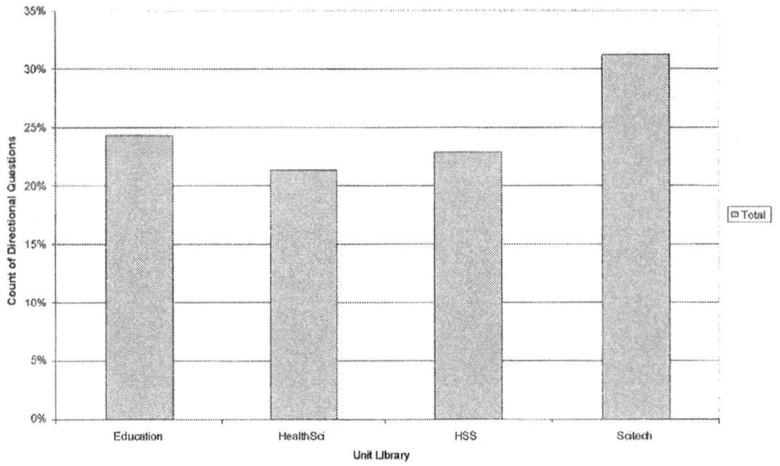


13

Fig. 6B: Types of Directional Questions by Large Unit Library

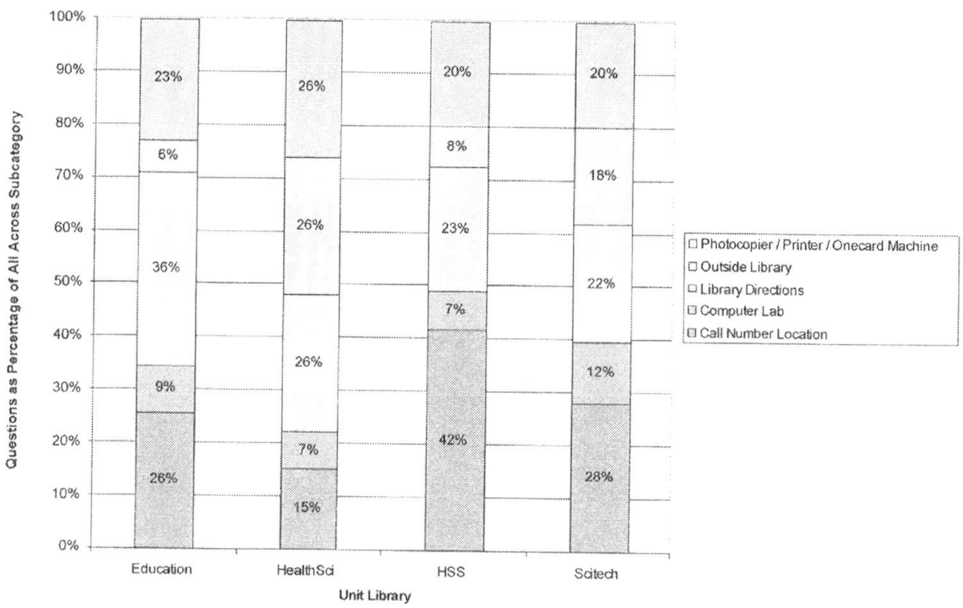


Regarding the medium / small unit library analysis, it should be noted that those libraries that received fewer than 10 directional questions were not included for analysis (Music, Math, Special Collections, and Data). As these libraries each have a specific focus of their collections or activity, they may be more "end destinations" and therefore less likely to receive directional questions.

The directional questions received by these libraries tended to focus on sites outside the library. Second most-frequently asked were questions asking about printing / photocopying or Onecard machines.

That the Knowledge Common (KC) was more likely to receive questions on the location of a computer lab likely speaks to the high demand for computers in this location.

14

Fig. 7A: Directional Questions by Medium / Small Unit Library
(n=2,041)

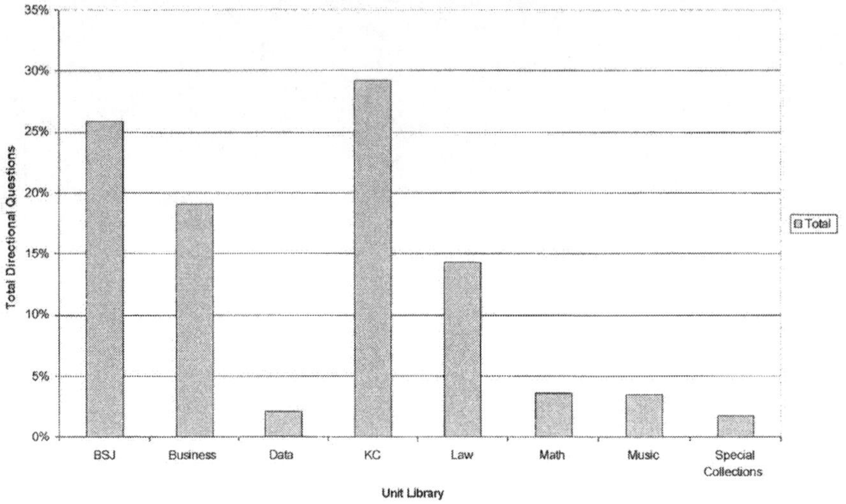


Fig. 7B: Types of Directional Questions by Medium Unit Library
(n=212)

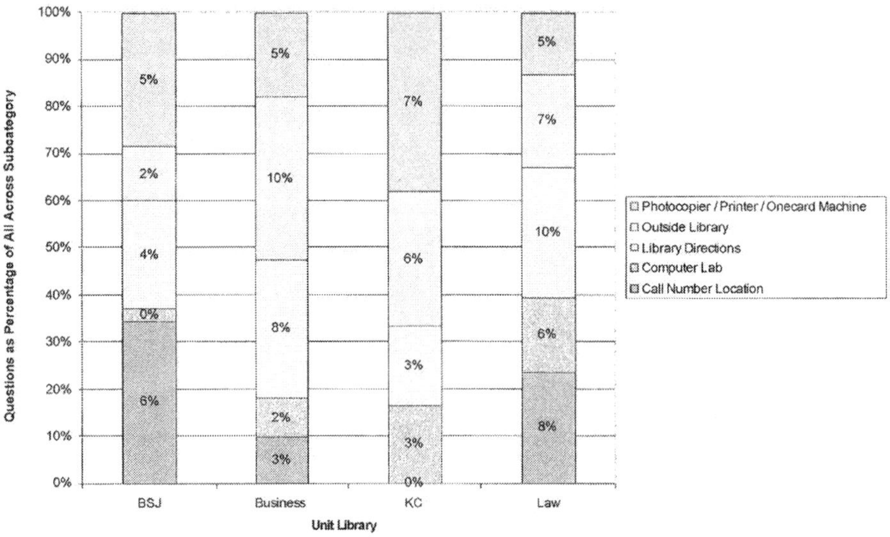


15

Other Thoughts/Observations:

Staffing Implications

While a single data collection during the first weeks of classes would not provide adequate information for staffing decisions, we can look at hourly questions received by the circulation and reference desks to see peak busy periods. For this time period, reference desks received the most questions between 12-3pm. Circulation desk questions were highest from 9am through 3pm, although these service points experienced much less of a spike than the reference desks.

Fig. 8: Questions Asked During Service Hours at Circulation and Reference Desks (n=6,737)

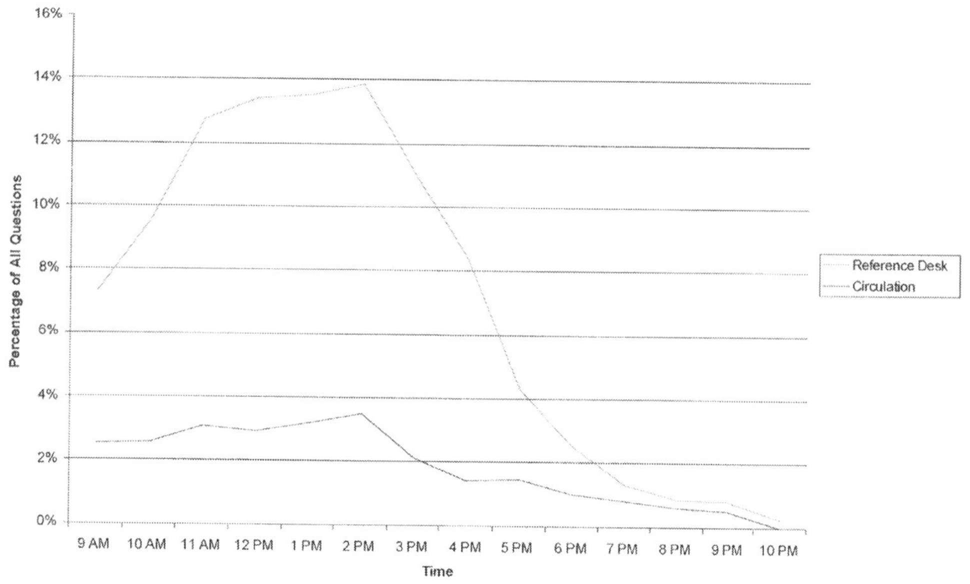


16

Thoughts on Data Gathering Methodology:

Regular statistics collection could be employed to fulfill the following additional purposes:

- Help understand the range of reference questions across the system
- Help inform reference training based on questions asked
- Provide general information to support staffing decisions (in conjunction with other data collection tools, such as Director's Station)

Some adjustments could be made to staff instructions, such as:

- Staff should note the specific question content, rather than using summary terms such as "directional"
- As well, there are instances where several questions were logged once, with notes on the number of times the question had been received (e.g., "5 x how to put money on Onecard"). For future collection purposes, it would be most effective to have staff ensure that each question was logged in separately, or to have the questions separated out during the categorization by the Assessment Librarian.

Any decisions made on the data collection timing and frequency should take into consideration the research questions to be answered, as well as the needs for facilitating staff participation and consistency of question logging.

Harold B. Lee Library
...BY STUDY AND ALSO BY FAITH

# Memo

**To:** Allyson Washburn, Web Working Group Chair
Scott Eldredge, Digital Imaging Manger

**From:** Brian Roberts, Process Improvement Specialist

**Date:** 11/10/2006

**Re:** Online Collections at BYU Survey Results

---

## INTRODUTION

To assess patrons' experience in using the utility developed through CONTENTdm to manage and display digital materials, a link to an online survey was placed on the home page of the Online Collections at BYU site, which serves as the gateway to BYU's digital library. Patrons that came to the Online Collections at BYU were invited to participate. This survey was intended to be part of an overall effort to assess the digital library at BYU, in terms of how it is named, how it is presented, and its functionality. Other studies are in process and reports will be done to summarize their findings. This memo summarizes the analysis and results of the Online Collections at BYU survey only.

## EXECUTIVE SUMMARY

The most striking thing to come from the survey was that those who responded were primarily coming to the Online Collections at BYU to view the Family History Archives (over 70%), which is explained by the status of those coming to the site, many being genealogists, family history researchers, or associated with the Family History Library or a local family history center. A majority of respondents visited the site more than once, although a good 36% of respondents were experiencing their first trip to the Online Collections at BYU. Interestingly, repeat visitors were more inclined to visit collections other than the Family History Archive than were first time visitors, implying over time they have become more familiar with what the digital library has to offer. First time users were generally less satisfied with their visit than repeat visitors, did not find the information as useful as repeaters, and rated it more difficult to find what they were looking for. Again, the obvious implication from this is over time, patrons became accustom to the utility and tended to be satisfied with its functionality and their ability to locate materials. And as expected, if a respondent was satisfied with the visit, they found what they were looking for and thought the process was easy. If they were not satisfied, they were not able to find the wanted information and found the process difficult. Respondents also commented that they liked its accessibility and the availability of the collections, and though they wanted to see more collections, many felt improvements could be made to the search capability and functionality of the utility.

## STUDY SPECIFICS & RESULTS

Over the course of several weeks, the survey to assess patrons' perceptions of their experience in Online Collections at BYU, the digital library of the Lee Library and Perry Special Collections, was made available to any and all who wished to participate as they visited the site and saw the survey link. In all nearly 450 patrons took the survey which yielded interesting results. Those that responded came from a wide range of possible respondents. Most came to the site via a DSL or Cable Internet connection, with 45 of the 50 states represented (most coming from either Utah or California). Most of the Canadian provinces were represented as were 8 foreign countries. And when queried as to their status, the vast majority indicated they were not from an academic setting. The majority indicated they were either a genealogist, family history researcher or associated with the Family History Library or a local family history center of the LDS Church. A summary of the demographics taken for this survey can be found in the set of charts below.

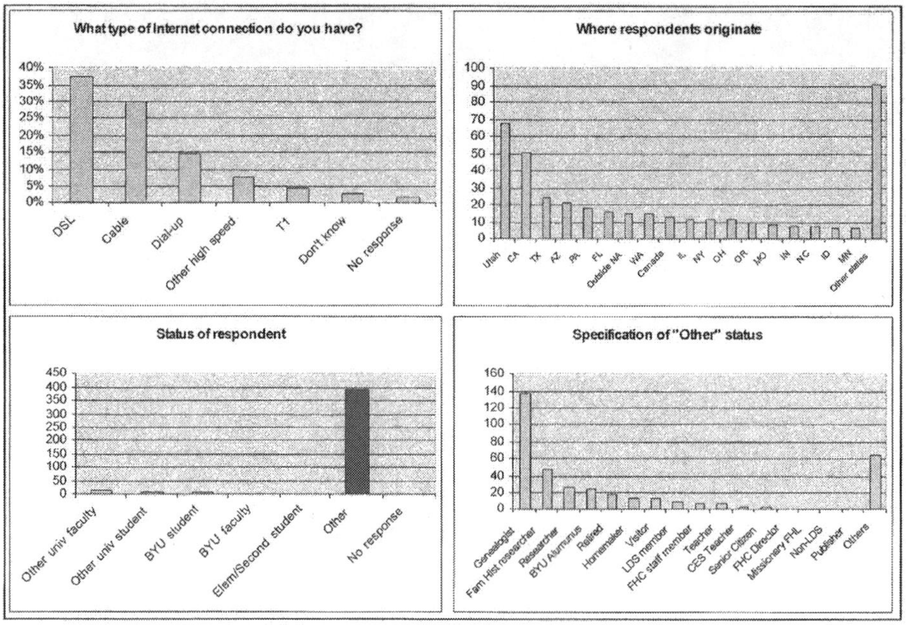


**Figure 1 - Demographics of respondents**

Though a majority of respondents said they had made multiple visits to the Online Collections at BYU (Daily, Weekly, Monthly or a few times a year), many indicated that this was their first time to the site (well over one-third). In addition, though many of the respondents indicated that they had discovered it through a friend's recommendation, a link from another site, a search engine, a link from the library catalog, or a newspaper article, most indicated some other means. These included mailing lists, family history centers, Rootsweb (a genealogical tool of Ancestry.com), genealogy newsletters, the Lee Library's home page, other genealogy periodicals, or through the Family History Library or LDS Church websites.

Based on those responses it was no surprise to see that the collection used most regularly was the Family History Archive. Over 70% of the respondents indicated that they came to the Online Collections at BYU to use that archive. But it was interesting to note how that was broken down by first time visitors and repeat visitors. It became very apparent that repeat visitors visited more of the various collections contained in the digital library than did the first timers, implying that over time users would become familiar with what Online Collections at BYU contained and would use the many collections available. This information has been summarized in the chart below.

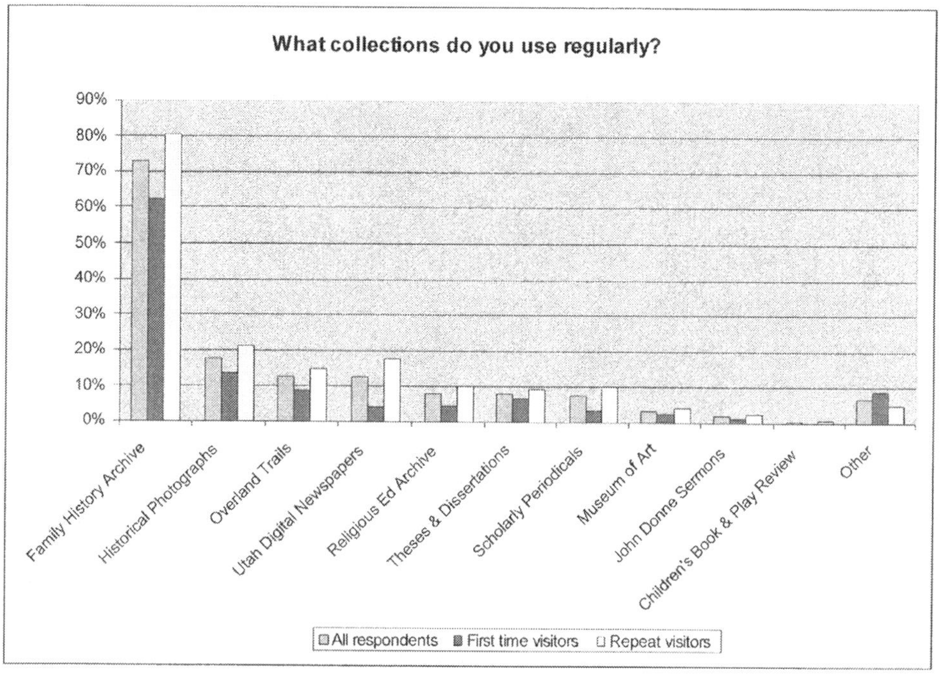


**Figure 2 - Collections used regularly by visitors to the Online Collections at BYU**

It was also interesting to note that nearly 60% of those that responded used only one collection in the digital library and of that, 84% used the Family History Archive. It is quite apparent that that collection appears to be the most valued among those frequenting the Online Collections at BYU.

When questioned about their overall satisfaction with their visit, the vast majority were satisfied (over 70%). However, again there was an interesting disparity in this between first timers and repeaters. Those that were experiencing their first trip to the Online Collections at BYU tended to be much more neutral or dissatisfied with their visit than were repeaters. This carried over into whether the information found was useful to them. First time visitors were less inclined to think so than were repeaters. First time visitors were also much harder on the utility than were repeaters when rating the ease in finding objects on the site. However, it should be pointed out that in all three instances, the majority of respondents, whether they were first time visitors or repeaters were satisfied with their visit, found the information useful and rated the utility easy in finding what they had come to search for. As stated earlier, the obvious implication from what these three questions are suggesting is that those that come to the site regularly can use it with ease, find what they are looking for, and in the end are satisfied with their experience. All three questions have been summarized and can be found in the charts below.

One final item to note, as cross comparisons in these three questions were made, one thing became quite obvious, if they were satisfied with their visit, then they had found what they were looking for to be useful and using the utility was easy. If they were not satisfied with their visit, they did not find the information useful and it was difficult to use. This should come as no surprise.

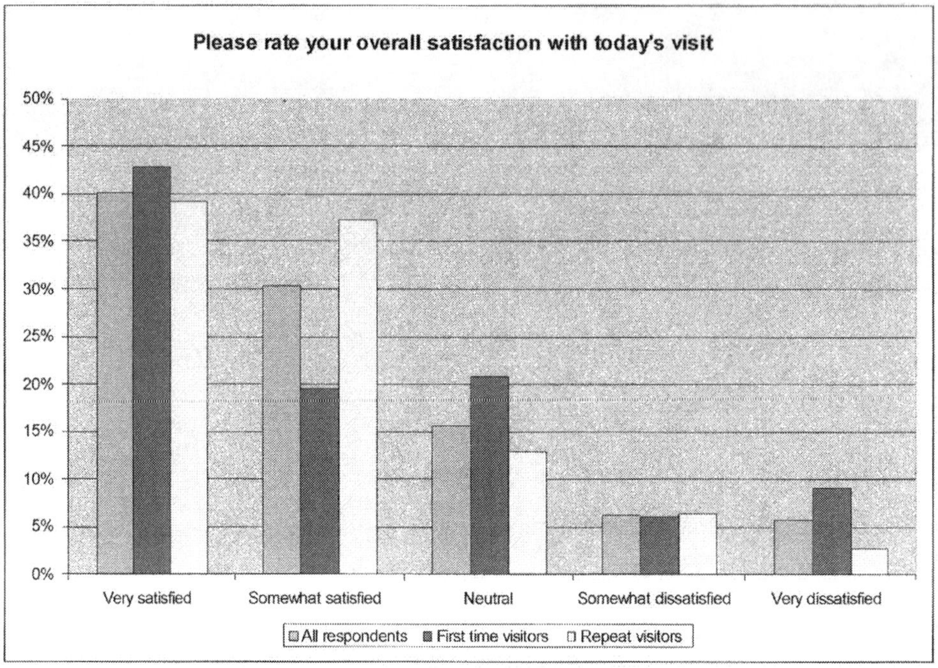
Please rate your overall satisfaction with today's visit
50%
45%
40%
35%
30%
25%
20%
15%
10%
5%
0%
Very satisfied
Somewhat satisfied
Neutral
Somewhat dissatisfied
Very dissatisfied
All respondents
First time visitors
Repeat visitors

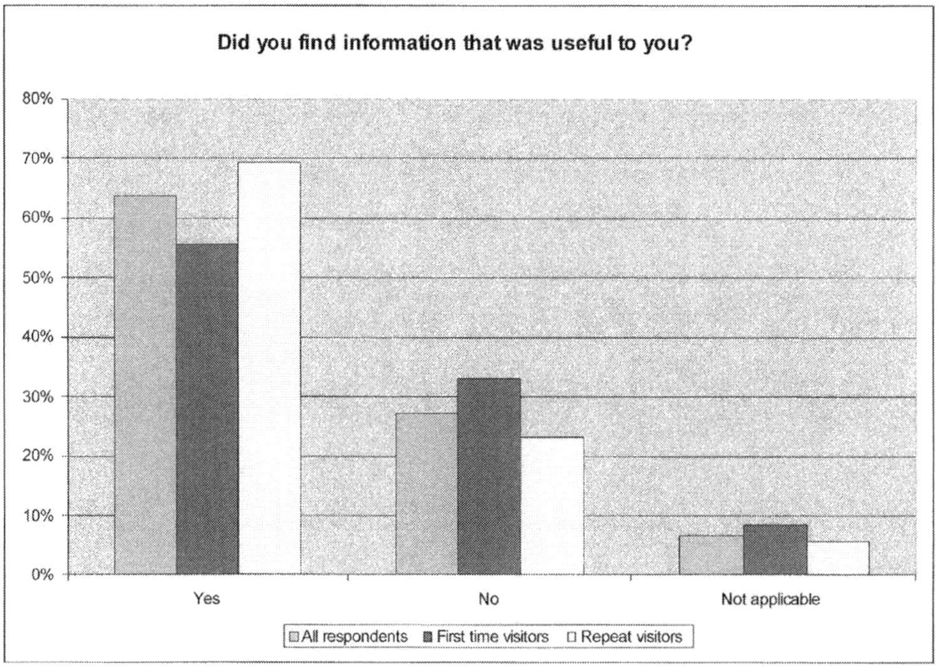
Did you find information that was useful to you?
80%
70%
60%
50%
40%
30%
20%
10%
0%
Yes
No
Not applicable
All respondents
First time visitors
Repeat visitors

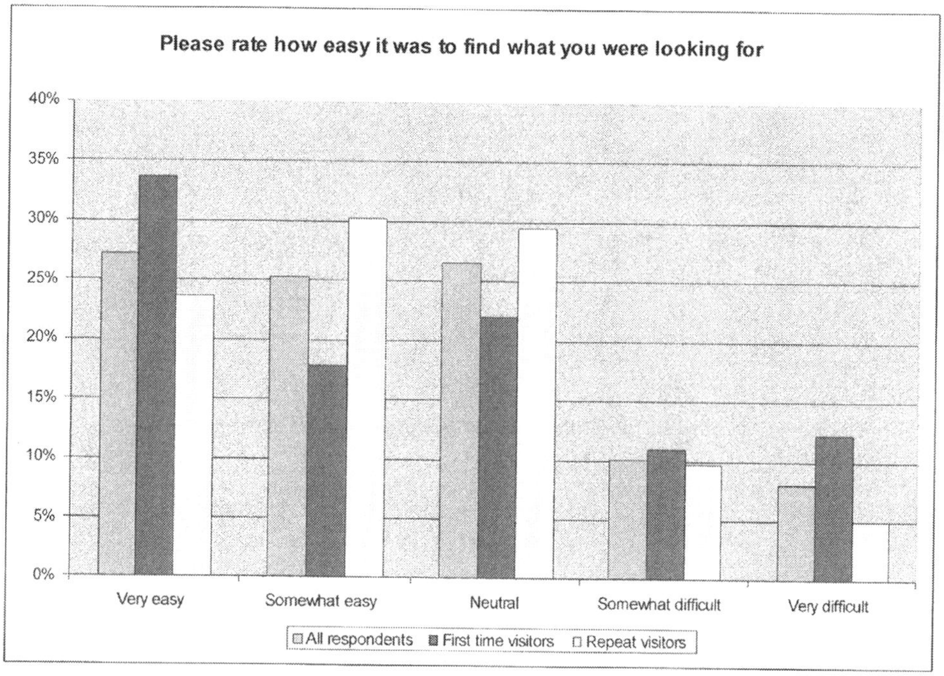


**COMMENT DATA**

Two additional questions were included at the end of the survey to elicit comments from respondents. In one they were asked what they liked about Online Collections at BYU, while the other asked what changes, if any, they would make to Online Collections at BYU. A total of 328 included comments to the first question, while 298 responded to the second (267 responded to both).

Overall, when asked what they liked about Online Collections at BYU, the responses were many and varied and most were quite positive. Respondents indicated they liked the online access, the availability of the collections, and the easy access to them. Others liked the search capability of the interface, enjoyed the vast amount of resources or thought the site was excellent. Any negative comment relative to this question tended to focus on the need of more materials in the collection (Mormon pioneer diaries or additional online books, for example). But they were few compared to the positive responses, which would be expected given the tenor of the question itself.

When asked what changes, if any, would respondents make to Online Collections at BYU, most said they wanted more resources (more histories, more rare books, more images, etc.). One-third of those that commented indicated such. But just over 40% indicated that searching could be improved (provide more options and better layout of results), that access could be simplified (stuff too buried or navigation cumbersome, for example), of general functionality of the application could be improved (PDF's hard to manage, OCR incomplete, etc.). There were several that had no comment because of a lack of experience using the Online Collections at BYU or simply thought it was fine the way it was. A few also indicated that the site could use clearer instructions, would like to see an index, or wanted a link that would show them recently added materials.

Comments from both questions have been summarized in the charts below.

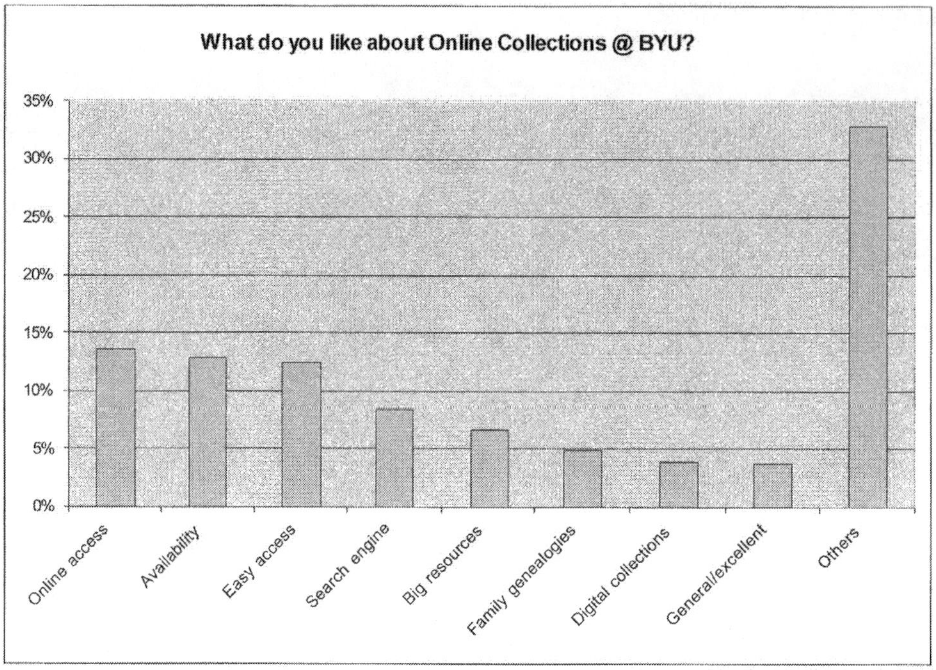
What do you like about Online Collections @ BYU?
35%
30%
25%
20%
15%
10%
5%
0%
Online access
Availability
Easy access
Search engine
Big resources
Family genealogies
Digital collections
General/excellent
Others

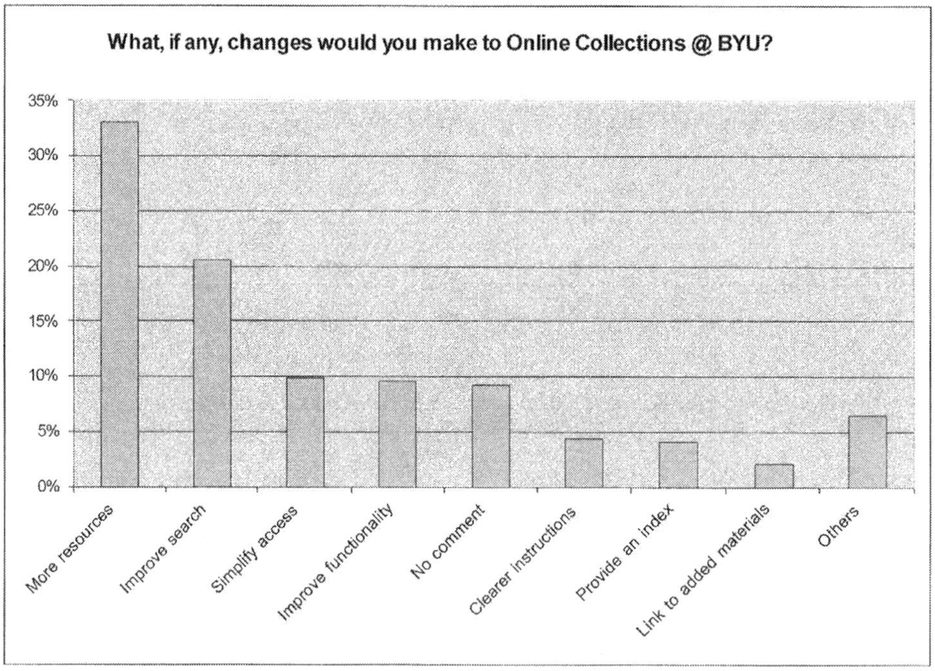
What, if any, changes would you make to Online Collections @ BYU?
35%
30%
25%
20%
15%
10%
5%
0%
More resources
Improve search
Simplify access
Improve functionality
No comment
Clearer instructions
Provide an index
Link to added materials
Others

## CONCLUSION

It is evident from the responses to the Online Collections at BYU survey that patrons like and use the service, particularly the Family History Archives. And though the users were varied in their backgrounds and from where they came, most tended to be genealogists or family history buffs, hence the heavy use of the Family History Archives. Overall, satisfaction of the utility tended to be high, although repeat users of the collections were more likely to be satisfied than first-time users, implying as familiarity increased with the site, so did their satisfaction level. And as expected, if they were satisfied with the site, they were more likely to have found what they were looking for and thought the utility easy to use.

And though there were many respondents that gave positive comments when asked what they liked about the site, most felt there could be steps made to improve the search capability of the utility, that efforts should be made to simplify access to the collections, and improvements were needed to the functionality of many aspects of the site.

It would seem that though the Online Collections at BYU is overall satisfying the needs of genealogists, family history buffs, and researchers with the content it contains, continued efforts should be made to continue to add content and make that content easier to access and use.

http://www.library.cornell.edu/staffweb/AnnualStatsArchive/Summary_ASR_6_6_07.pdf

http://www.library.cornell.edu/staffweb/AnnualStatsArchive/Summary_ASR_6_6_07.pdf

# CORNELL UNIVERSITY LIBRARY ANNUAL STATISTICS 2005/2006

The Cornell University Library's annual statistical report is prepared by the PSA Research & Assessment Unit. Please direct questions, comments or requests for copies to:

Research & Assessment Unit
Cornell University
215 Olin Library
Ithaca, New York 14853-5301

Telephone: (607) 255-7401
Fax: (607) 255-2493

February 2007

http://www.library.cornell.edu/staffweb/AnnualStatsArchive/Summary_ASR_6_6_07.pdf

CORNELL UNIVERSITY LIBRARY – ANNUAL STATISTICS 2005/2006

**July 1, 2005 – June 30, 2006**

**TABLE OF CONTENTS**



iii

http://www.library.cornell.edu/staffweb/AnnualStatsArchive/Summary_ASR_6_6_07.pdf

## REPORT VERSIONS AND HIGHLIGHTS

### Report versions available

This report is available in 2 versions: this summary version and a full version. Both are available in PDF at http://www.library.cornell.edu/staffweb/AnnualStatsArchive/indexStatArch.html. The summary version includes the highlights, as well as the summary tables and graphs from the full version. Detailed figures and notes for individual units are available in the full version, which, for the first time, provides links to tables and graphs from its table of contents.

### Collections

- The number of print volumes increased by 121,929 or 1.6% (full version Table 1a). This is the smallest number of gross volumes added in the last 15 years, in part reflecting rising costs and the shift to electronic formats.
- The number of physical serial subscriptions decreased by 1,916 or 3.1% (full version Table 3). This reflects, in large part, the effort underway to eliminate unnecessary and costly duplication of serials in physical and electronic formats. This is the fifth consecutive year the number of non-electronic subscriptions and commercial binds has decreased (Summary Graphs 1 & 5).
- At the same time, the number of e-journal holdings in the Library Catalog increased by 6,068 or 14.7% (full version Table 4).
- In the fall of 2005, the Library began a project to move close to 1 million volumes to the Library Annex. With the net addition of 374,178 volumes in 2006, the Library Annex holds over 2.2 million volumes or 28.6% of the Cornell Library's print collection (full version Table 1).
- The number of slides and filmstrips increased by 12,050 items or 30.4%, reflecting, in large part, a transfer of slides from University Photography to RMC.

### Services/Usage

- The number of circulation transactions for library materials decreased by 1.5%, while the circulation of equipment purchased for patrons' use in the Library (ranging from laptops for use within the Library, to polling devices used by students in their classes) increased by 11.8% (Summary Table 2). The number of charges decreased by 2.9%, while renewals increased by 4.1%. Total circulation (of all collections and equipment) showed a 0.1% decrease.
- The volume of Interlibrary Loans (ILL) continued to increase this year. The lending showed a lower rate of increase (1.9%) than borrowing (7%) (Summary Table 2). Similar to last year, Borrow Direct, a service that allows patrons to directly initiate expedited requests after a combined search of the library catalogs of Brown, Columbia, Cornell, Dartmouth, University of Pennsylvania, Princeton and Yale accounted for 33.3% of filled lending and 44.9% of filled borrowing requests respectively (full version Table 8).
- Library-to-library delivery is a service that allows Cornell-affiliated patrons to request that a book be sent from one library to another for more convenient pickup. Jointly, the libraries that reported in 2005 saw a 33.1% increase in 2006 (full version Table 6a).

iv

http://www.library.cornell.edu/staffweb/AnnualStatsArchive/Summary_ASR_6_6_07.pdf

- The use of the Library's Enhanced Returns policy, implemented in 2001, which allows patrons to return items borrowed to any library, increased by 7.6%; 16.1% of total returns in 2006 were made at libraries other than the owning library (full version Table 5b).
- In October 2005 , the Library launched a new service to deliver books to faculty members' departmental offices. The Library filled 1,280 such requests from October through June (full version Table 7).
- 17,139 people (or an average of over 100 people per night) used Uris Library between 2 and 8 a.m. during the academic year when Uris is open for study 24 hours, Sunday-Thursday (full version Table 16 footnote).
- The number of user instruction sessions offered and session participants were up 22.2% and 6.4% respectively (Summary Table 2).
- The use of the citation management software, Refworks, and the journal table of content delivery service, MyContents, are shown in Table 10. This year, users added 351,237 references (compared to 99,959 for half of 2005) to RefWorks through library databases interfaces as well as manually. The use of MyContents increased, but at a slower pace. At the end of 2006, 1,934 users (a 2.1% increase from 2005) received 10,340 tables of contents electronically from 959 journal titles.
- The use of the Library Gateway increased by 23.5% from last year: close to 9 million visits to the Library Gateway were tracked by Web analysis software (full version Table 9b).

We welcome your comments and questions about this report.

The Research and Assessment Unit

http://www.library.cornell.edu/staffweb/AnnualStatsArchive/Summary_ASR_6_6_07.pdf

# SIZE OF COLLECTIONS

## SUMMARY TABLE A
### SIZE OF COLLECTIONS OVER TIME
2000 - 2006

| Item Count | 2000 | 2001 | 2002 | 2003 | 2004 | 2005 | 2006 | Total Growth (2005 - 2006) Incr/(Decr) | Percent Change (2005 - 2006) Incr/(Decr) |
|---|---|---|---|---|---|---|---|---|---|
| Printed Volumes & Non-Book Materials (Tables 1 & 2a) | | | | | | | | | |
| Printed Volumes | 6,830,411 | 6,975,415 [1] | 7,139,192 [1] | 7,316,826 [1] | 7,477,388 [1] | 7,586,799 [1] | 7,708,728 | 121,929 | 1.6% |
| Maps | 237,066 | 239,526 | 241,553 | 244,556 | 246,881 | 249,000 | 250,050 | 1,050 | 0.4% |
| Motion Pictures | 3,932 | 4,039 | 4,044 | 4,157 | 4,360 [1] | 4,360 | 4,419 | 59 | 1.4% |
| Filmstrips and Slides | 40,370 | 40,981 | 43,362 | 40,251 | 37,027 [1] | 39,627 | 51,677 | 12,050 | 30.4% [2] |
| Video Tapes and DVDs | 14,623 | 15,052 | 18,530 [1] | 20,372 | 22,343 | 23,709 [1] | 26,335 | 2,626 | 11.1% |
| Sound Recordings | 90,015 | 90,395 | 104,515 | 105,643 | 112,005 | 113,901 | 115,765 | 1,864 | 1.6% |
| Computer Files | 9,070 | 9,569 | 11,484 | 12,819 | 15,213 [1] | 16,510 | 17,871 | 1,361 | 8.2% |
| Microforms (Tables 2b) | | | | | | | | | |
| Microfilm | 196,622 | 199,732 | 204,396 | 207,715 | 210,775 | 214,883 [1] | 217,526 | 2,643 | 1.2% |
| Microfiche | 6,264,099 | 6,413,284 | 6,557,254 | 6,596,881 | 6,706,067 | 6,829,128 [1] | 6,905,399 | 76,271 | 1.1% |
| Microcards | 139,494 | 139,456 | 139,456 | 139,456 | 169,484 [1] | 169,484 | 169,484 | 0 | 0.0% |
| Microprints | 1,048,409 | 1,048,409 | 1,048,409 | 1,048,409 | 1,048,409 | 1,048,409 | 1,048,409 | 0 | 0.0% |
| Archival & manuscript materials (in cu. ft.) [3] (Table 2c) | 64,509 | 65,131 [1] | 65,839 | 67,020 | 68,299 | 69,565 | 70,271 | 706 | 1.0% |
| Physical Serial Subscriptions (Table 3) | 63,306 | 64,891 | 64,559 | 63,870 [1] | 62,937 | 62,494 | 60,578 | (1,916) | (3.1%) |
| Physical Newspaper Subscriptions (Table 3) | 492 | 492 | 479 | 374 | 345 | 368 | 370 | 2 | 0.5% |
| Networked Electronic Resources [4] (Table 4) | 4,625 | 9,312 | 40,888 | 107,892 | 150,504 | 375,236 [1] | 394,474 | 19,238 | 5.1% |
| Full-text journals | 3,321 | 5,347 | 19,616 | 20,241 | 29,159 | 41,237 [1] | 47,305 | 6,068 | 14.7% |
| Other | 1,304 | 3,965 | 21,272 | 87,651 | 121,345 | 333,999 | 347,169 | 13,170 | 3.9% |

**Notes:**

[1] Corrected/adjusted count.

[2] Increase reflects in large part the transfer of slides from University Photography.

[3] Figures for the Medical Archives are in linear feet.

[4] Does not include resources exclusive to the Medical College.

1

http://www.library.cornell.edu/staffweb/AnnualStatsArchive/Summary_ASR_6_6_07.pdf

## SIZE OF COLLECTIONS

### SUMMARY GRAPHS 1-2

PRINTED VOLUMES, NON-BOOK, MICROFORM, MANUSCRIPT AND PRINT SERIAL SUBSCRIPTION TRENDS, 2000-2006

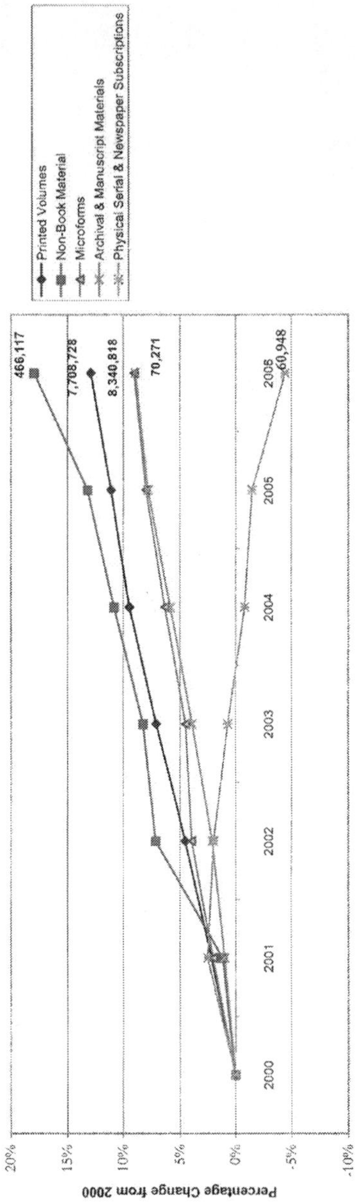


NETWORKED ELECTRONIC RESOURCE TRENDS, 2000-2006

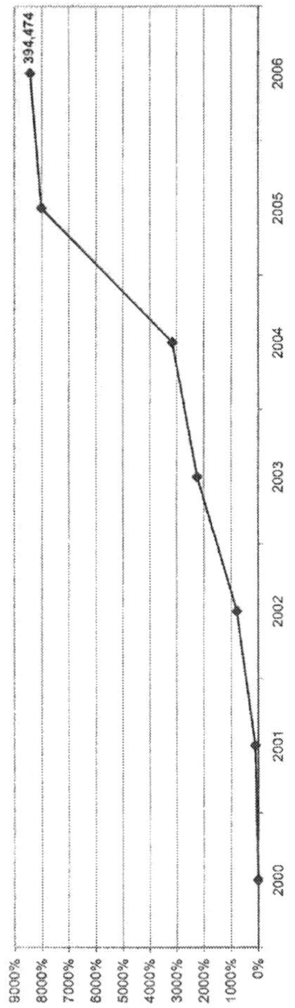


2

http://metrics.library.upenn.edu/FACTS06.pdf

University of Pennsylvania Library

PennLibrary FACTS

2005|2006

http://metrics.library.upenn.edu/FACTS06.pdf

# PennLibrary FACTS

2005 | 2006

The data in this publication describe the development and use of Penn Library resources from July 1, 2005 through June 30, 2006

## Table of Contents



Access this publication online through the Penn Library Data Farm at http://metrics.library.upenn.edu/prototype/datafarm. For information about Library measurement, contact Joe Zucca, Management Information Services and Communication, 215-573-4643/zucca@pobox.upenn.edu

University *of* Pennsylvania Library | www.library.upenn.edu

**Executive Summary**

LibQual survey scores for 2006 show steady progress over the past three years in satisfaction ratings of USC Libraries among students and faculty. 2006 scores were higher in every dimension (affect of service, information control, and library as place), and were consistently higher in nearly every subcategory. Along with the core survey questions, scores were also higher in every instance for the supplemental questions relating to general satisfaction (such as "service quality" and "how I am treated") and information literacy outcomes (such as "helps me stay abreast in my field", and "aids my academic advancement").

The representation for 2006 maintained the pattern of past surveys, where Liberal Arts and Sciences are somewhat overrepresented, and where Viterbi and Marshall are underrepresented. As such, data may not be completely applicable to the Business Library or the Science & Engineering Interdisciplinary Center. Both Marshall and Viterbi did respond, but not in proportion to their size. The highest response came from social sciences departments within the College. Among user groups, graduate students responded more than faculty or undergraduates. Since undergraduates typically score the library higher than other user groups, the overall score would be expected to drop for this year. The higher overall scores for this year, therefore, demonstrate a significant improvement over the past three years.

In comparing USC's 748 survey responses to the 99,155 responses from other academic libraries within the Association for Research Libraries, USC scores fall slightly below the perceived mean in all three dimensions. In the Information Control and Library as Place dimensions, however, USC's desired expectation was higher than the ARL mean. Higher expectations from the USC community on issues within these areas could partly explain the lower scores.

This year's survey shows no change in the past 3 years in the trend of library use. Students and faculty prefer to obtain information online, yet at the same time the data demonstrates a continuing interest in library spaces. Qualitative data shows a high demand for both group and individual study spaces, as well as comfortable spaces with stable wireless access.

Finally, although service quality scores have improved, some service issues still exist. Users expect highly skilled and knowledgeable library personnel who are eager to meet their needs. Better online system usability and online help is also a frequent request. Based on the scores and the comments, however, the most important issue for users is the collection. The lowest scores appear in the categories of "print and/or electronic journals I require for my work" (Information Control question 8), and "printed library materials I need for my work" (Information Control question 5). Overall, 2006 data supports a need to increase collection development and outreach efforts, while continuing to improve service quality through training, and continuing to creatively redesign library spaces.

**Next Steps**

Scores and comments will be reviewed by the Leadership Team with an eye toward the FY08 budget. Key issues with budget implications will be targeted and addressed. In addition, all library teams, centers, and departments will review LibQual quantitative data along with the qualitative data (reported separately) and discuss library and university-wide implications. Upon review, team leaders, center directors, and library managers are expected to develop action plans for quick wins within their respective areas. Suggestions for policy or process changes, or remedies with multi-year budget implications, will be forwarded to the Leadership Team for approval before being implemented.

**Sections**

http://www.lib.virginia.edu/mis/benchmarking/bench-ProcessRept.html

UNIVERSITY of VIRGINIA LIBRARY Management Information Services

- About MIS
- Balanced Scorecard
- Surveys & Reports
  - Library Surveys
  - Benchmarking Team Reports
- Statistics
- Web Usability Testing
- Links

# Benchmarking Team

## Report on Benchmarking Process

### January-May 1999

Benchmarking is an ongoing, systematic process for measuring and comparing the work processes of one organization to those of others that exhibit functional "best practices." The goal is to provide an external standard for measuring the quality and cost of internal processes, and to help identify where there may be opportunities for improvement. To be effective, benchmarking should be integrated into operations throughout the organization and should be an ongoing process that analyzes data collected over time. It is a learning process that helps institutions discover how they can best improve the services, direct or indirect, that they offer to their customers.

### The Charge

For the 1998/2000 biennium, the University of Virginia Library chose as one of its goals (6f) to institute benchmarking as a tool for the analysis of internal processes and to establish benchmarks against which the Library can measure those processes. The Library's first Benchmarking Team was created in January 1999. The Team was charged with two challenges: to create a benchmarking process for the Library; and to carry out a short-term benchmarking project as a pilot (for which there is a separate Shelving Report). The benchmarking portion of the charge included:

". . . The Team is charged with learning the benchmarking process and applying it to a specific project. The intent is http://www.lib.virginia.edu/mis/benchmarking/bench-ProcessRept.html

that the members of this Team become the Library's core staff with knowledge of benchmarking. After learning the process, the Team members should be able to:

- assist other groups with their benchmarking projects
- assist in developing benchmarking expertise among other staff members, for example, by participating in a training program

Each May the membership of the Team will be reviewed. Those who want to remain on the Team will be joined by new members so that the Benchmarking Team can be a constantly renewed central group of experts in the process. New projects will be determined at the same time that membership is reviewed." (see Appendix 1 for full charge.)

The benchmarking pilot project was chosen by User Services and Central Services Councils after review and discussion of the results of several user satisfaction surveys conducted in the spring of 1998. It was decided that our reshelving process was limited enough for the Team to use as a pilot for learning the process.

## Team Members

Team members were chosen by Management Information Services staff and Kendon Stubbs, and were selected in part from members who had similar experience on other process improvement teams. It was also important to have representation from several departments and service units affected by the project. Two Team members were from Management Information Services to provide statistical skills and continuity for the benchmarking process. The Team consisted of David Griles from Management Information Services, Doug Moseley from Cataloging, Heather Packard from Science/Engineering, Gary Treadway from Social Sciences Services, and Lynda White from Fine Arts/Management Information Services. Two Team members from stacks supervisory staff were added within a few weeks: Don McCracken, Stacks Supervisor in Alderman, and Pam Howie, Public Services Library Assistant in Music.

## The Learning Curve

The Team began its task by identifying and reading books and articles on benchmarking in industry and the military. There is some literature on benchmarking specifically relating to libraries, but details on how to carry out the process in libraries are generally lacking. In addition, it could not be determined that there is any training available locally through the University. No courses are taught through Organizational Development and Training, through the Commerce School, the Education School, nor through the Darden Business School. Inquiries to Association of Research Libraries went unanswered. The Training Coordinator for the University Library bravely stepped in and began educating herself on the process. She was, of course, on the same learning curve as the Benchmarking Team, making it difficult to develop a timely class for the team. A query to the LARGE_PSD listserv, asking for contact with those who had done a benchmarking project, brought a response from Pennsylvania State University's Sally Kalin. She graciously consented to spend some time on the phone explaining the process and also to send a packet of information on the benchmarking projects she had participated in. In addition, the Team spent some time reading several books and articles on benchmarking. Fortunately, after a short time, the litany of benchmarking became repetitive and the Team decided to embark on its pilot project.

The basic benchmarking process is straightforward (see Appendix 2 for greater detail):

1. Determine what to benchmark
2. Form a benchmarking team
3. Identify benchmark partners
4. Collect and analyze benchmarking information
5. Take action

## Collecting Data

The Team undertook several parts of the process simultaneously. Since there were minimal statistics or other data available on our shelving process, we began to flowchart the process in all 11 libraries and to work on a survey instrument to help us gather data about the process as practiced at the University of Virginia Library. The questionnaire was tested by interviewing stacks supervisors in units where all returned items were not shelved by the end of each day. The outcome was messy at best. It was necessary to revise the questionnaire several times in order to achieve more consistent answers.

## Best Practices

While the Team was brainstorming questions for our internal survey, we also began to explore how to identify best shelving practices at other institutions. The literature on the shelving process is a sparse as the literature on benchmarking in libraries. Instead of relying on the literature, two electronic listservs (LARGE_PSD and CollDev) were queried with the assistance of Diane Walker and Gary Treadway. Those responding to the listserv query were initially asked whether they would be willing to participate in a brief survey. The 19 institutions that responded were sent a short 10-question survey (Appendix 3) devised to ferret out best practices at institutions similar to the University of Virginia Library. Thirteen institutions responded over the next two months revealing much interesting data about shelving standards and staff sizes. From these responses the Team was able to identify several institutions having what appeared to be "best practices." Contacts made with American Library Association's Library Administration and Management Association officers revealed that no LAMA committee members were aware of institutions doing either benchmarking or shelving studies.

## Consultant

The conversation with Sally Kalin of Pennsylvania State University about benchmarking led us to invite Gloriana St. Clair of Carnegie-Mellon University in Pittsburgh to the University of Virginia. Ms. St. Clair presented basic benchmarking information to the entire Library staff, and she assisted the Team in revising the local practices questionnaire and in deciding which institutions exhibited "best practices." She also suggested that the Team was moving toward its objective at a good pace in spite of its reservations about the lack of training in the benchmarking process. She confirmed that the Team should stop reading and "just get on with it."

After Ms. St. Clair's visit, the Team made rapid progress revising the local questionnaire (see Appendix 4). Answers garnered in the initial staff interviews were re-entered in the revised document and were much clarified in that process.

## Measurements

The Team concurrently began to devise a plan to measure several things for which there was no data: how fast books are shelved (books per hour), what the turnaround time is (from return desk to shelf), how accurately books are shelved, and what the turnaround time is for pick-ups. David Griles developed the protocol and ran the Sirsi reports with which the studies were done. With the exception of Science/Engineering, Team members carried out the measurements in libraries other than their home libraries.

## Site Visits

Simultaneously, Team members began planning for site visits to the University of Arizona and Virginia Tech. These two institutions were chosen because of their reports of 4-hour turnaround time, 94%+ accuracy rates, and previously completed shelving studies. The site visits were planned for mid-April at the same time most of the Team was measuring speed, turnaround time, accuracy, and pick-ups. The site visits were essential for understanding how the best practices really worked. There is no substitute for walking through a process and having an opportunity to ask questions along the way. In addition, the host libraries were asked to fill out the same survey that had been completed by our own stacks staff. This allowed us to identify procedures that were similar and different, thus pointing to how our process could be improved.

## Communication

At various points in the project, the Team apprised staff and stakeholders of progress by:

- having stakeholders on the team
- making direct contact with other stacks supervisors
- inviting Ms. St. Clair to present information on benchmarking to the entire staff
- sending an email interim report mid-way through the project to Library@Virginia.edu. (See Appendix 5.)

## Report

Using and comparing data from the questionnaire, the best practices email survey, the site visit reports, and our own local measurements, the Team was able to develop recommendations for changes in the shelving process at the University of Virginia. A report on the project, with these recommendations for action, was submitted to the Library's Administrative Council.

http://lib.washington.edu/usability/newDesign04/Fall2004FinalReport.pdf

# UW Libraries Usability Testing Report: New Design Fall 2004

November 14th, 2004

I

http://lib.washington.edu/usability/newDesign04/Fall2004FinalReport.pdf

## Table of Contents

# 1. Executive Summary

This report documents the findings of usability testing performed fall 2004 of the newly designed UW Libraries web site. Ten UW Library users with a range of library web site use performed tasks determined to help understand the ability of users to access resources using the libraries website.

The usability testing indicated that finding resources using the newly designed Libraries website is generally fairly easy. Many of the tasks were completed quickly and easily by participants.

Users easily found:

- Purchase requests
- Library hours
- Citations from the homepage
- The book renewal web page
- Course reserves (via myuw)
- Journals from the catalog in the e-journals page

However, users had difficulties:

- Knowing what fell under "more" on the homepage
- Distinguishing between 'major' and 'complete' databases
- Recalling introductory information provided on the resource pages
- Find services provided by departments within the library
- Understanding the need to search the catalog using the journal title, not the article title

The site received compliments from every usability test participant. These include: "I wish we had this twenty years earlier," "I like having a catalogue link on the home page", and "I love the libraries."

# Assessment Plans

# ASSESSMENT PLAN

Columbia University Libraries

*2007 through 2009*

**February 1, 2007**

**CUL Assessment Team**
Jennifer Rutner
Bill Sees
Jeff Carroll
Joanna DiPasquale
Jane Winland
Iris Wolley

assessment@columbia.edu

## Executive Summary

In 2006 an *ad hoc* Assessment Team was gathered and charged with exploring assessment activities across Columbia University Libraries. The Team examined assessment as a concept and practice in libraries, conducted interviews with staff members, worked to define what assessment means for CUL, outlined assessment goals and priorities, and developed the recommendations in this Plan.

Goals of an assessment program at CUL:

- To foster a culture of assessment at the Libraries.
- To enable data-driven decision-making at all staff levels.
- To promote information transparency in the work environment.

*(see page 8 for further explanation)*

## Recommendations

**Assessment Priorities**

The following areas have been identified as needing assessment support, and will be given priority. *(see page 11 for further explanation)*

Ongoing
- Usability Program
- Reference Services Assessment
- Collection Development analysis support

2007
- Library as Place: Construction and Renovation Projects
  - Lehman Library Renovation
  - Information Commons: Lehman, Butler, Sciences
  - Business Library Renovation
  - The Sciences Library
  - Geology Library Renovation
  - Starr Library Renovation
- Faculty Brochure Survey
- Access Services Quality Service Survey
- Document Delivery and Paging Services, planning support
- Faculty Focus Groups

2008
- Library as Place: Construction and Renovation Projects
  - Avery Library
  - The New Business Library
  - Math Library
- Graduate Student Focus Groups

2009
- LibQual+ Survey

- o Undergraduate Focus Groups

**Creation of an Assessment Working Group (AWG)** *(refer to page 7 for further explanation)*
The formation of an ongoing Assessment Working Group (AWG) is recommended. The charge of the AWG includes initiating, supporting and monitoring a sustainable assessment program at CUL, assisting in identifying and maintaining assessment priorities, ensuring that appropriate training and support for staff is provided, publishing results of assessment initiatives as appropriate and advocating for a culture of assessment at the Libraries.

The Director of Access Services will serve as the liaison between the Assessment Working Group and the Management Committee.

**Online Assessment Center (SWIFT)** *(refer to page 16 for further explanation)*
The Assessment Center will provide staff with direct access to training opportunities, resources on assessment, guidelines for assessment initiatives, contact with the AWG, and updates on assessment projects.

Long-term goals for the Assessment Center include data bank-like functionalities, enabling staff to access information and run reports independently.

https://www1.columbia.edu/sec/cu/libraries/staffweb/adm/assessment/index.html

3

The Assessment Coordinator would like to thank the members of the Assessment Team for their contributions to the development of the Assessment Plan. Their various perspectives, expertise and experience were invaluable.

The Assessment Team, an *ad hoc* group, was gathered to support the Assessment Coordinator in the research and preparation of this plan.

**Membership:**
Access Services: Bill Sees
Collection Development: Jeff Carroll
LDPD: Joanna DiPasquale
Reference: Jane Winland
Technical Services: Iris Wolley

## Introduction

Assessment encompasses all areas of information gathering (qualitative and quantitative) and analysis in relation to decision-making and the management of the Libraries. Collections analysis, usability studies, user input studies, and internal analysis of policies and procedures are currently the four major areas of assessment at CUL. Assessment attempts to measure CUL's collections, websites, services, policies and procedures against the users' expectations and CUL's strategic goals and mission. Assessment initiatives must have a compelling motivation, a rigorous and valid research process, and a commitment to engaging and applying the information gathered.

CUL is an organization that is familiar with assessment. In recent years the Libraries have conducted two rounds of LibQual+ Surveys, User Input Focus Groups, the E-Reference Assessment, numerous usability tests, staff focus groups, and "local" surveys, as well as continually collecting data for ARL Statistics. Reference staff found the analysis of E-Reference transactions to be enlightening, especially in determining staffing models for future E-Reference services. In general, staff have found that opportunities to gather and interpret user feedback have proven to be incredibly valuable and informative. These efforts have illustrated user needs that may not have been previously perceived, and have allowed the Libraries to focus on improving service for user. Library use statistics such as gate counts have enabled the Business and Social Sciences libraries to advocate for extending service hours. All in all, CUL has a stable, well-established foundation of conducting assessment that can now be expanded, enhanced, and formalized through the implementation of the Assessment Plan.

An Assessment Program is intended to provide staff with the tools necessary to make the decisions that are expected of them throughout the continuous process of managing and working in the Libraries. The following is an example from Jane Winland, the Director of Social Sciences Libraries at Lehman Library:

> *Assessment has helped me considerably in getting funding. We had to make a decision on how to prioritize our renovation project; having data from students on their needs for the library space and services allowed me to advocate for implementing those changes.*

In addition, assessment aims to support and enable staff to improve in all aspects of service:

> *Transforming our libraries to reflect a culture of assessment is essential to increasing our success with customers and stakeholders, and maintaining relevancy in a competitive environment.* (Lakos, Amos and Shelley Phipps. "Creating a Culture of Assessment: A Catalyst for Organizational

Change," Libraries and the Academy, Vol. 4, No. 3, pp. 345-361. The Johns Hopkins University Press, Baltimore, MD, 2004.)

A successful assessment program depends upon the following:

- A focus on supporting the patron's information and communication needs
- Performance measures that are included in the strategic plan
- A commitment to supporting assessment from the Libraries' leadership
- Staff recognition of the value of assessment
- Staff support for and participation in assessment initiatives
- Continuous communication with patrons
- Routine collection and analysis of data and user feedback
- Commitment to applying assessment results in decision-making

Derived from: *Amos Lakos, " Defining a Culture of Assessment"*
*http://personal.anderson.ucla.edu/amos.lakos/CUtoolkit.html*

In an effort to better understand the current state of assessment at the Libraries, the heads of each library, as well as the key players in Technical, Public, and Access Services were interviewed. The following concerns and needs were identified:

**CUL concerns:**

- Not enough time and support for analyzing and engaging data collected in assessment efforts
- Limited ability to access data independently
- The need to make statistics gathering routine and consistent
- Targeting data gathering towards information needs and strategic goals
- Organizational commitment to data-driven decision-making is not established: Is this a priority? Expectation?
- Maintaining the ability to tailor assessment tools to a library's unique needs
- Lack of transparency about and lack of access to current data sets
- "Survey Fatigue"

**CUL needs:**

- To make assessment initiatives, such as statistics-gathering and focus groups, routine
- The ability to freely access data independently, immediately
- To develop tools and skills supporting the design and implementation of assessment initiatives
- Support for data analysis and displaying data effectively
- A central location for the storage and sharing of data
- Training, workshops, best-practices for assessment

6

## Assessment Working Group

The formation of an ongoing Assessment Working Group (AWG) is recommended. The charge of the AWG includes initiating, supporting and monitoring a sustainable assessment program at CUL, assisting in identifying and maintaining assessment priorities, ensuring that appropriate training and support for staff is provided, publishing results of assessment initiatives as appropriate and advocating for a culture of assessment at the Libraries.

The first priority of the AWG is to create an Implementation Plan for tracking the progress of the outlined assessment priorities. Project management software to support these efforts is being sought. The Implementation Plan will outline the goals, timeframe, key players, and methods of assessment for each project, and will assign an AWG member to work with the key players and the Assessment Coordinator to ensure that the assessment is carried out successfully.

For example, the Faculty Brochure Survey goals are to understand which library services faculty are aware of, which services they value the most, and about which services they would like to learn more. Key players include the Communications Committee and selectors. The information is needed by the end of February 2007 in order to stay on-task with the brochure project timeline. This information will be gathered via an online survey for timeliness, ability to document the data collection, and convenience for the participant. This data, in combination with information gathered via the LibQual+ Survey and the User Input Focus Groups will inform decisions regarding marketing services to faculty. The Communications Committee will develop, administer, and monitor the survey and is committed to applying the information collected to their project.

The AWG will work closely with the Management Committee to ensure that strategic planning initiatives have the appropriate assessment support, and that the information needs of the Management Committee are satisfied in a timely manner. The Director of Access Services will serve as the liaison between the Assessment Working Group and the Management Committee.

In this manner, the AWG will function as a consulting group for assessment initiatives. Staff members designated as Strategic Plan project leads are encouraged to contact the AWG regarding assessment projects.

**Membership**

Assessment Coordinator (Chair)
Director of Access Services (Core Member)
A.D. of Collection Development (Core Member)
Web Services Librarian, or appropriate counterpart from LDPD (Core Member)
Access Services Committee Member (rotating, two year commitment)

7

Reference Coordinating Committee Member (rotating, two year commitment)
2 members (two year commitment)

Manager of Training Programs (*Ex Officio* Member)

The Assessment Team proposes a nomination process to fill the remaining seats on the AWG. Professional staff who are interested in joining the AWG and supporting assessment initiatives at the Libraries are asked to self-nominate or nominate their colleagues. Nominations can be sent to jb2563@columbia.edu. The AWG will review the nominations and select 2-3 staff members, in consultation with supervisors, to serve on the Working Group. Every effort will be made to ensure that all divisions of the Libraries are equally represented through the membership and work of the AWG.

The AWG is committed to building relationships with partners on campus and in libraries across the country, and will explore exciting new ways to develop such partnerships.

## Goals & Priorities

**Culture of Assessment**

CUL is committed to building a culture of assessment within the organization.

> "A Culture of Assessment is an organizational environment in which decisions are based on facts, research and analysis, and where services are planned and delivered in ways that maximize positive outcomes and impacts for customers and stakeholders. A Culture of Assessment exists in organizations where staff care to know what results they produce and how those results relate to customers' expectations. Organizational mission, values, structures, and systems support behavior that is performance and learning focused."
>
> – *Amos Lakos*
> *http://personal.anderson.ucla.edu/amos.lakos/*

**Data-driven Decision-making**

One goal of building a "culture of assessment" at the Libraries is to foster a deeper commitment toward incorporating data into the decision-making process of all levels of staff. Amos Lakos states that "In essence, we are advocating a change in our institutional cultures, from a static, institutionally and professionally inwardly focused culture to an externally focused institution and profession that needs to embrace the notion of decision-making based on measurements, and analysis based on customer expectations." (Evidence Based Library Management – A View to the Future, pg 3.) Data-driven decision-making

8

is a concept closely related to evidence-based management. "It just means finding the best evidence that you can, facing those facts, and acting on those facts – rather than doing what everyone else does, what you have always done, or what you thought was true." (http://www.evidence-basedmanagement.com) CUL is moving towards creating a work environment where staff are encouraged and expected to base decisions on information gathered to support and inform our management needs. Of course, not every decision requires a thorough data-analysis. The goal here is to identify information needs, and perform the necessary assessment, enabling data-driven decision-making.

One great example of data-driven decision-making involves determining terminology for a website. This type of decision is clearly an opportunity to gather information from users that will directly inform the decision. The website is a service that the Libraries provides for users. It is a simple exercise to put together a list of possible terms and consult users to discover which ones have the most meaning for them. Collection development is another area where data can directly inform decision-making. Even renovation projects give opportunity to consult library users: Should we get couches or armchairs? Carpet or tile? These are all opportunities for staff to step back, ask questions, and collect relevant information from users to guide decisions and enable CUL to provide the best possible service.

**Information Transparency**

Ensuring that all staff have access to organizational information, as appropriate, is a major priority of the assessment program. "Information transparency" refers to the end result of breaking down internal barriers that prevent access to valuable information. This can mean providing fluid and immediate access to meeting minutes, circulation statistics, gate counts, shelving statistics, survey results, etc., to all members of the staff. In each case, there is currently someone at CUL who is responsible for this information. This often means that staff are required to consult this person for access to the information. Having a culture of "information transparency" would mean that staff are able to *independently* access all types of organizational information without having to involve an intermediary.

The Assessment Center on SWIFT will be tool allowing staff to access information independently, and in timely manner.

One long-term goal for the Assessment Center includes data-bank-like functionalities, allowing staff to access reference statistics, circulations statistics, gate counts, etc., from all libraries through one central interface. (Much like the Penn Data Farm.) This type of tool contributes greatly to a "information

9

transparency", and a "culture of assessment" in that staff are enabled to access necessary information, when they need it, and run reports independently – helping to make assessment an easy, natural, part of working at CUL.

**Training**

In coordination with the Manager for Staff Training, topical training, workshops, and information sessions will be held to prepare staff for participation in developing and administering assessment initiatives. Staff education topics include focus group facilitation, interview skills, survey development, project management, change management, and techniques for displaying data effectively.

**Data Analysis**

A major concern among CUL staff is time and the ability to analyze and apply data collected from an assessment effort. The Assessment Coordinator and the AWG will provide support for data analysis and interpretation, as well as assistance developing charts and graphs. Staff members can request support by contacting the AWG or the Assessment Coordinator.

Critical analysis of complex data requires some skills that are not typical among library professionals. Most library assessment literature recommends seeking partners in the larger community who can help with advanced statistical data analysis. The AWG is committed to forging relationships at the University that can supply this type of support.

**Assessment Program Priorities**

Based on reviewing the 2006-2009 CUL Strategic Plan, User Input Focus Group results, and the 2006 LibQual+ Survey, the following assessment priorities have been identified. These initiatives will be given priority in terms of resources (funding and staff time), support, and staffing over the next three years, and are subject to change with the environment.

- **Usability Program**
  The Strategic Plan implicated usability in many areas of online tool development, such as "seek[ing] out or develop[ing] tools that use catalog data to effectively aid resource discovery and access," "develop[ing] a common interface and request service for the ReCAP and collections," and "reviv[ing] the idea for a Special Collections Gateway." The EPAG group works diligently to review E-resources pages, request usability studies and suggest improvements. In 2006 the LibQual+ Survey indicated that the Libraries were not meeting user expectations regarding "a library website that enables [users] to locate information on [their] own." Other areas of the LibQual+ Survey also implicated usability, though often

10

indirectly. Faculty assigned low scores for "print and/or electronic journal collections [they] require for [their] work" - indicating a possible deficiency in marketing, and technological barriers preventing user-friendly access to the sometimes complex resources CUL provides. In order to support CUL's commitment to developing relevant, intuitive online tools and websites for users, LSO and LDPD are dedicated to incorporating standard usability practices to analyze those online tools. A comprehensive usability program will:

- Establish routine usability procedures for LDPD and LSO, integrated into the development of online tools and websites
- Establish a method for staff and committees to request usability tests for existing websites and tools
- Provide a pool of participants for usability tests, as well as incentives
- Support and maintain prioritization of usability projects
- Explore partnerships with CNMTL, DKV and CU regarding usability practices
- Explore concepts of information architecture, and how this will play a role in website re-design and management

Usability Project Priorities:

- LSO
    - Spring 2007
        - E-journals websites
        - Library Public Workstation Re-design
    - Future projects
        - Course Reserves
        - New Books Lists
        - E-journal analysis: Partner with ILL and Collection Development
        - SWIFT
        - Voyager: Partner with peer institutions using Voyager to discover and develop interface design best-practices
- LDPD
    - Spring 2007
        - RBML and Archives websites
        - FAQ development (in coordination with Access Services)
        - HR websites
    - Future projects
        - "Behind the Scenes" websites
        - "Request It" websites

11

- "Services" websites
- Digital Collections and Special Collections websites

- **Library As Place: Construction and Renovation**

The following building projects were identified in the Strategic Plan, and will incorporate user input as to the configuration of the spaces, as appropriate. Methods of incorporating user input into construction and renovation planning include surveys, focus groups, discussion groups, and observational studies. The AWG will work with the lead staff member of each project to ensure that assessment needs are met.

  - Information Commons: Lehman, Butler, Sciences
  - Lehman Library Renovation
  - The Sciences Libraries
    - build a new partially consolidated science library
    - renovate the Geology Library space
    - refurbish or renovate the Math/Science Library
  - Business Library
  - Avery Library
  - Starr Library

- **Reference Services**
  - Indirectly mediated
    - FAQ development (in coordination with Access Services)
    - Subject Guides
  - Directly mediated
    - Understand use patterns, preferences and motivation for using Reference services via the desk, website, email, IM or Chat and other online tools. Gauge user satisfaction, and identify user expectations and needs
    - E-Reference Analysis
      - Gather user input on e-reference tools
      - Determine preferred methods of communication
      - Understand expectations and needs for reference support
      - Gauge user satisfaction with current E-Reference options; provide a benchmark for upcoming changes
  - Develop methods for benchmarking and measuring all new initiatives in Reference Services
  - Track the use of print and online reference resources to better understand use and inform collection development
  - Quality Service

12

- Participate in Quality Service assessment initiatives
- Assess Reference transactions for quality and user satisfaction

- **Quality Service Continual Survey**

The Access Services Division will be piloting a Quality Service Continual Survey in Butler Library for the Fall of 2007. This survey will enable users to provide feedback on any Access service point in Butler Library (Circulation, LIO, ILL, PMRR, and the Stacks), via online forms. Feedback will be collected in a database, and will be accessible by staff for analysis.

This survey will be structured so as to be easily scalable and incorporate Access service points at all Libraries, as well as Reference service points. The AWG will be working with the Access Managers and LSO on the development and marketing of the survey.

- **Document Delivery and Paging Services**

The AWG will work with the Director of Access Services and the Head of ILL to establish assessment projects in support of these initiatives.

- Paging Service:
  - Benchmarking Questionnaire
  - Statistics gathering
  - Cost-benefit-analysis: "book-by-mail" vs. CUL Mail Service models
  - Develop system requirements for a tracking system
  - Faculty Survey
- Document Delivery:
  - Benchmarking Questionnaire
  - Process Analysis
  - Interface Usability Studies

- **Collection Development**

The following assessment priorities were identified by the Director and A.D. of Collection Development. Assessment methods will be identified to support the following projects:

- Gather and analyze e-resources use statistics
  - Explore software tools to support these efforts
- Update Collection Development policies
- Explore the liaison experience, in relation to collection development
- Develop a documentation system for the vendor analysis process

13

  - Analysis of Borrow Direct / ILL statistics, partner with LSO
  - Explore software tools for collection analysis

- **LibQual+ Survey**
  - CUL will continue to use the ARL LibQual+ Survey to gauge user perceptions of our performance as a "big picture" marker.
  - LibQual+ will be used to track our progress in key areas of concern, as well as identify future assessment needs.
  - LibQual+ will be used on a three year cycle; the next instance being Spring of 2009. The schedule is easily adaptable if circumstances change.
  - CUL will make an effort to identify peer institutions who also participate in the LibQual+ Survey, and encourage coordination and sharing results in an effort to provide benchmarking.

- **Annual Focus Groups**
  - In an effort to build avenues of communication with users, provide continual feedback, support ongoing strategic planning efforts and build on the success of the User Input Group, the AWG will coordinate annual focus groups of user populations. Based on a three year rotation, aligned with the LibQual+ survey schedule, focus groups will be held for the core user populations: faculty, graduate and undergraduate students.
    - 2007: Faculty
    - 2008: Graduate Students
    - 2009: Undergraduate Students
  - Topics for the focus groups will be determined based on key areas of concern identified through the LibQual+ results, strategic goals, and consultation with the Public Services Committee.

*A note on ARL Statistics and the Libraries Manager of Financial Planning*: The Assessment Coordinator will keep close contact with Jane Sahner, and seek new ways to apply and publish data collected annually for ARL Statistics, ensure that data collection mechanisms are relevant and usable, and provide unfettered access to these data and reports.

*A note on incentives:* A budget has been established to fund incentives for users participating in assessment efforts. To request incentives for an assessment initiative, please contact the Assessment Coordinator. Special thanks to the Deputy University Librarian for supporting this valuable initiative.

14

## Roles

### Assessment Coordinator

The Assessment Coordinator (i.e. Program Coordinator for Marketing and Assessment), reporting to the Director of Access Services, functions as an initiator, coordinator and advocate for assessment at CUL. The Assessment Coordinator is kept aware of all assessment initiatives taking place in the Libraries, provides support and direction as needed, chairs the Assessment Working Group and manages assessment priorities, ensuring that CUL's information needs are met.

### Management Committee & Strategic Planning

As the Assessment Plan and priorities are based on the direction of the CUL Strategic Plan, and functions to support strategic planning goals, communication between the AWG and the Management Committee will be critical to the success of the Assessment Plan.

Leadership support is an indicator of a strong culture of assessment. The AWG, Assessment Coordinator, and the Assessment Plan's success depend heavily on the support and involvement of Library Directors in assessment initiatives and their advocating for and participating actively in building a culture of assessment at the Libraries.

### Requesting Assessment

Staff members who wish to request an assessment initiative, or seek advice or support for assessment in their libraries are invited to contact the Assessment Coordinator. Depending on the scope of the assessment initiative in question, the Assessment Coordinator will be able to fulfill the request directly, or will refer the request on to the AWG, who will review requests, and determine their level of priority. Priority will depend upon relation to the strategic goals and staff availability for necessary support. The AWG will work with the requester to explore the topic at hand, and develop a plan for the assessment initiative.

Please notify the Assessment Coordinator of any assessment initiatives in the Libraries in an effort to maintain information transparency and coordination across the system. Coordination is one way of ensuring the avoidance of survey fatigue.

## Online Assessment Center – SWIFT

The Assessment Center will connect staff directly with assessment tools, best practices, and valuable data. The Assessment Center will also serve as the main pathway for communication between staff, the AWG and the Assessment

15

Coordinator. Staff will be able to track the progress of assessment initiatives, request assessment projects, request support or training, and access data independently.

https://www1.columbia.edu/sec/cu/libraries/staffweb/adm/assessment/index.html

**Data Bank**

A long-term goal for the Assessment Center is to house a CUL Data Bank that would facilitate access to statistics that staff regularly require for their work. Examples include: reference statistics, collection use statistics, e-resources use statistics, CLIO statistics, etc.

The planning and execution of the Data Bank will be managed by the AWG, in conjunction with LSO's database programmer, and LDPD.

## The Future of Assessment at CUL

**User-centered Design**

The AWG is committed to exploring opportunities for incorporating methods of ethnographic research and observational studies as an element of a user-centered design approach to the development of services and online tools. User-centered design aims to enable the organization to better understand user behaviors, needs and expectations regarding information gathering.

**Learning Outcomes**

The AWG will investigate methods for measuring learning outcomes in relation to bibliographic instruction and general library use, in an effort to understand and measure the influence of the Libraries on scholarship at Columbia University.

**Disseminating Assessment Results**

The AWG will explore avenues for communicating assessment results to the Libraries' staff, patrons, administration, and colleagues around the country. Publishing the results of assessment efforts, and illustrates the Libraries commitment to employing the information collected, our commitment to our users and our value of their input.

**CUL Annual Reports**
The AWG strongly encourages Library Heads and Directors to include a review of assessment initiatives in each library's annual report.

**Assessment Forums**
The AWG will arrange for regular Assessment Forums to be held at the Libraries. Topics should address various methods of assessment, and will contribute to the building a culture of assessment here at CUL. Potential topics include: planning assessment initiatives, focus groups, assessment and marketing: a dynamic relationship, assessment success stories, Penn Data Farm, decision-making, and change management.

**Identifying Partnerships**
The Assessment Coordinator and the AWG are dedicated to identifying potential partners in assessment, on campus as well as at peer institutions. The AWG will assist in reaching out to appropriate partners at CU, and building outside relationships.

**Assessment "Fellow"**
Each year the AWG will nominate one staff member who has shown considerable interest in assessment, initiative in supporting assessment, and has become an advocate for a culture of assessment in the Libraries. This person will be recognized by the Libraries and given opportunity to develop his or her assessment skills by either attending a conference, workshop, or training session on an assessment topic, at the Administrative rate.

## Results of the CUL Culture of Assessment IQ Test

In January 2007, a "Culture of Assessment IQ Test" was distributed as a survey to professional staff. About 30 staff responded. This tool, based on the work of Amos Lakos, served as a benchmark for gauging CUL's understanding of assessment, by definition, and the extent to which the concept and practice of assessment are present in the organizational culture.

The results of the survey, as illustrated below, illustrate the *majority* response for each question.

| | |
|---|---|
| Assessment is evident in our library planning documents such as strategic plans | True |
| Assessment is a campus priority | False |
| Library administrators are committed to supporting assessment | True |
| Staff accepts responsibility for assessment activities | True |
| There is support and rewards for staff who engage in assessment | False |
| Library policies and procedures are designed to enable, not inhibit, user information needs | True |
| Collaboration and cooperation exists among individuals and departments of the Libraries | True |
| My library considers user needs when allocating resources | True |
| My library actively cultivates a positive relationship with its users | True |
| Assessment leads to results in my library | False |
| My library routinely collects, uses, and disseminates meaningful user data and feedback | False |
| My library evaluates its operations and programs for quality | True |
| Staff have expertise and skills in assessment | False |

These results, combined with the responses to final question "What does 'assessment' mean to you?," clearly indicate that CUL has a clear understanding of the concept of assessment, and values assessment. Assessment is present in the strategic plan, the perception that library administration is committed to assessment, and staffs' expressed willingness to accept responsibility for assessment are all signs that there is a strong foundation for a more comprehensive assessment program to build on.

There are, of course, challenges with assessment as well. While the concept of assessment is clearly understood and valued, the practice of applying assessment data to management practices has not yet become part of our culture. False responses to "assessment leads to results in my library," "my library routinely collects, uses, and disseminates meaningful user data and feedback," and "staff has expertise and skills in assessment" illustrate areas for improvement.

## Bibliography

Beck, Susan J. Making Informed Decisions: The Implications of Assessment, April 10-13, 2003: ACRL Eleventh National Conference. Charlotte, North Carolina: American Library Association, 2003.

Bentley, Stella and Bill Myers. "Creating a User-Centered Culture of Assessment" (PowerPoint Presentation). 31 Jan. 2007 <http://www.informationservices.ku.edu/assessment/presentations/CreatingCultureOfAssessment.ppt>

Hiller, Steve. "Assessing User Needs, Satisfaction, and Library Performance at the University of Washington Libraries." Library Trends 49 (2001): 605-625.

Hiller, Steve and James Self. "From Measurement to Management: Using Data Wisely for Planning and Decision-Making." Library Trends, 53 (2004): 129-155.

Lakos, Amos and Shelley Phipps. (2004). "Creating a Culture of Assessment: A Catalyst for Organizational Change." Portal: Libraries and the Academy, 4 (2004): 345-361.

Lakos, Amos. "Evidence Based Library Management - A View to the Future." Amos Lakos Homepage. 31 Jan. 2007 <http://personal.anderson.ucla.edu/amos.lakos/>

Library Assessment Blog 31 Jan. 2007 <http://www.libraryassessment.info>.

Norlin, Elaina. Usability Testing for Library Websites. Chicago: American Library Association, 2002.

Penn Data Farm 31 Jan. 2007 <http://metrics.library.upenn.edu/prototype/datafarm/>

19

http://www.howard.edu/library/assessmentplan/main/GoalsandObjectives.htm

Library System

HOURS | CONTACT US

HOWARD UNIVERSITY

HOME | SEARCH / BROWSE | RESEARCH ASSISTANCE | SERVICES | SPECIAL PROGRAMS | ABOUT US    SITE MAP

HUL HOME > Assessment >

Library Assessment Plan

## GOALS AND OBJECTIVES

*Related pages >* Survey Questionnaire | Values & Beliefs, Vision, Mission

**GOAL 1:** *ACCESS TO INFORMATION -- Patrons informational needs for education, research and service are fully met*

- **Objective A:** > Template
  ***Patrons have access to resources that adequately support their information needs, regardless of format, time, or location***
- **Objective B:** > Template
  ***Patrons use high performance technologies to locate the information they need, regardless of format, time, or location***
- **Objective C:** > Template
  ***Patrons obtain information not available at Howard libraries, promptly and with minimum effort***

**GOAL 2:** *INFORMATION EMPOWERMENT -- The student's information skills as foundations for life-long learning are developed or enhanced*

- **Objective A:** > Template
  ***Patrons receive effective ready reference and consultation assistance*** (Satisfaction)
- **Objective B:** > Template
  ***First-year students can use finding tools effectively to identify useful resources and locate needed information*** (learning outcome)
- **Objective C:** > Template
  ***Upper-level undergraduates and graduate students completing departmental research methods courses can readily identify appropriate finding tools expeditiously, evaluate information sources and their content critically, and document information content accurately*** (learning outcome)
- **Objective D:** > Template
  ***Students and faculty will acquire advanced skills in searching particular finding tools--Sterling Online Catalog; WWW; specialized databases*** (learning outcome)

**GOAL 3:** *APPROPRIATE ENVIRONMENT -- Presentation and exchange of knowledge and ideas are facilitated by appropriate spaces*

- **Objective:** > Template
  ***Patrons will experience an inviting, safe, secure and resourceful environment that nurtures creative thinking, productivity and intellectual fulfillment*** (Satisfaction)

**GOAL 4:** *EFFECTIVE ORGANIZATION -- Enhanced organizational effectiveness and efficiency, and sustained financial growth*

- **Objective A:** > Template
  ***Worksforce is effective, informed and motivated***
- **Objective B:** > Template
  ***Library has a financial management strategy***

HOME | SEARCH / BROWSE | RESEARCH ASSISTANCE | SERVICES | SPECIAL    SITE MAP

http://www.lib.utexas.edu/vprovost/assessment/pdf/UTlib-Assessment-Plan_07182006.pdf

The University of Texas at Austin
Administrative Unit Assessment Plan

University of Texas Libraries
Fred Heath
June 9, 2006

Updated July 19, 2006



## I. Mission Statement

The University of Texas Libraries advances the academic mission of the University and enriches the intellectual life of the people of Texas by fostering information discovery, enabling teaching and research, nurturing creativity, partnering in the development and dissemination of new knowledge, and contributing to the intellectual growth and fulfillment of the individual.

## II. Goals

**1. Access and services**
Deliver the broadest array of information resources, systems and services, accessible on demand, to enable users to satisfy their information needs.

**2. Leadership**
Provide library leadership that results in effective service provision and efficient management and operations.

**3. Spaces**
Create library spaces that enable and support evolving learning and research behaviors by providing comfortable and functional learning environments.

http://www.lib.utexas.edu/vprovost/assessment/pdf/UTlib-Assessment-Plan_07182006.pdf

## III. Objectives

**1. Access and services**

a. Build our collections of record by adding key resources.

b. Provide users with immediate access to information at the point of need by supplying more electronic information through collaborative initiatives and by transforming spending patterns.

c. Organize the libraries around products and services most valued by our users, and retrain and realign staff to deliver those products and services more effectively.

**2. Leadership**

a. Develop a sustainable business model for library operations, collections, and spaces.

b. Put the advancement plan (attached) into effect, including activities of the advisory council, fundraising efforts, and collaboration with others on campus.

c. Advance collaboration with University of Texas System libraries, other research libraries in Texas, and other state and regional institutions and organizations.

**3. Spaces**

a. Re-engineer library public spaces to accommodate current and emerging technologies and user behaviors.

b. Improve user seating and group study spaces within library facilities.

c. Redesign public areas in the Libraries to enable patrons to use spaces and services more productively; make the spaces more attractive to users and potential donors.

http://www.lib.utexas.edu/vprovost/assessment/pdf/UTlib-Assessment-Plan_07182006.pdf

## V. Implementation Plan

| Objective | Measure(s) of success | Assessment Method | Responsible Party | Task | Timetable |
|---|---|---|---|---|---|
| **Access and Services** | | | | | |
| Build our collections of record by adding key resources. | University of Texas Libraries is ranked by the Association of Research Libraries as one of the top fifteen ARL Libraries based on collections | Association of Research Libraries survey | Assessment Coordinator and Associate Director for Research Services | Collect unit statistics (includes gate count, circulation, reference transactions) | Due first week of the month |
| | | | | Submit unit statistics to ARL | Completed annually in the fall |
| | | | | Analyze and report on results | Survey results are disseminated by ARL one year after submission |
| | UT Libraries maintain a satisfaction rating in the LibQUAL+ survey in the category of Information Control that is equal to or higher than its peers | LibQUAL+ Survey | Assessment Coordinator and Associate Director for Research Services | Run LibQUAL+ survey | Annually/Biennially in spring semester |
| | | | | Analyze LibQUAL+ survey results (qualitative and quantitative) | Spring and summer of survey year |
| | | | | Report on LibQUAL+ results | Fall and winter of survey year |
| Provide users with immediate access to information at the point of need by supplying more electronic information through collaborative initiatives and by transforming spending patterns. | Steady shift in resource allocation from print materials to electronic materials | Budget analysis of expenditures | Executive Associate Director and Associate Director for Research Services | Review Libraries budget and assign funding | Annually |
| | Steady increase in downloads | Usage statistics for Libraries web site and online resources | Assessment Coordinator and Associate Director for Research Services | Analyze Libraries usage statistics and vendor provided usage data and report to Administrative Council | Annually |

University of Texas Libraries Assessment Plan
7/19/06

10

http://www.lib.utexas.edu/vprovost/assessment/pdf/UTlib-Assessment-Plan_07182006.pdf

| Objective | Measure(s) of success | Assessment Method | Responsible Party | Task | Timetable |
|---|---|---|---|---|---|
| Organize the libraries around products and services most valued by our users, and retrain and realign staff to deliver those products and services more effectively. | Analysis of trend data pulled from the LibQUAL+ survey shows customer satisfaction improving year to year | LibQUAL+ Survey | Assessment Coordinator | Run LibQUAL+ survey | Annually/Biennially in spring semester |
| | Continued use of Libraries facilities and services | Collection of monthly unit statistics | Assessment Coordinator, Associate Directors for Public Services, Student Services | Collect unit statistics (includes gate count, circulation, reference transactions, web site usage) | Due first week of the month |
| | | | | Organize and analyze unit statistics | Biannually |
| | | | | Report on unit statistics (including posting online for Libraries staff use) | Annually |
| | Libraries staff spend 5% of work year in training and development | Libraries training database | Head of Libraries Human Resources | Report on LibQUAL+ results | Fall and winter of survey year |
| | | | | Run annual report on staff training time | Annually |
| | | | | Analyze and report on results | Survey results are disseminated by ARL one year after submission |
| | UT Libraries maintain | LibQUAL+ Survey | Assessment | Run LibQUAL+ survey | Annually/Biennially in |

University of Texas Libraries Assessment Plan
7/19/06

11

http://www.lib.utexas.edu/vprovost/assessment/pdf/UTlib-Assessment-Plan_07182006.pdf

| Objective | Measure(s) of success | Assessment Method | Responsible Party | Task | Timetable |
|---|---|---|---|---|---|
| | an overall customer satisfaction rating in the LibQUAL+ survey that is equal to or higher than its peers | | Coordinator | | spring semester |
| | | | | Analyze LibQUAL+ survey results (qualitative and quantitative) | Spring and summer of survey year |
| | | | | Report on LibQUAL+ results | Fall and winter of survey year |
| **Leadership** | | | | | |
| Develop a sustainable business model for library operations, collections, and spaces | University of Texas Libraries is ranked by ARL as one of the top fifteen ARL Libraries overall | Association of Research Libraries survey | Assessment Coordinator | Collect unit statistics (includes gate count, circulation, reference transactions) | Due first week of the month |
| | | | | Submit unit statistics to ARL | Completed annually in the fall |
| | | | | Analyze and report on results | Survey results are disseminated by ARL one year after submission |
| | UT Libraries maintain an overall customer satisfaction rating in the LibQUAL+ survey that is equal to or higher than its peers | LibQUAL+ Survey | Assessment Coordinator | Run LibQUAL+ survey | Annually/Biennially in spring semester |
| | | | | Analyze LibQUAL+ survey results (qualitative and quantitative) | Spring and summer of survey year |
| | | | | Report on LibQUAL+ results | Fall and winter of survey year |
| Put the advancement plan | Relative standing of | Monthly Gift Reports | Vice Provost, Chief | Review reports | Monthly |

http://www.lib.utexas.edu/vprovost/assessment/pdf/UTlib-Assessment-Plan_07182006.pdf

| Objective | Measure(s) of success | Assessment Method | Responsible Party | Task | Timetable |
|---|---|---|---|---|---|
| into effect, including activities of the advisory council, fundraising efforts, and collaboration with others on campus. | Libraries endowments steadily improves when measured against ARL Libraries and UT Austin campus units | | Development Officer, Major Gifts Officer | distributed by the Office of the Vice President for Development | |
| | | Libraries Advancement Plan | Vice Provost, Chief Development Officer, Major Gifts Officer | Review and update advancement plan | Annually |
| Advance collaboration with University of Texas System libraries, other research libraries in Texas, and other state and regional institutions and organizations. | Increasing investment in resources by UT System and improving cost avoidance strategies | Budget analysis of expenditures | Executive Associate Director, Associate Director for Research Services | Review Libraries budget and consortial purchases and assign funding | Annually |
| | Steady increase in downloads within UT System | Usage statistics for UT System and other consortial resources | Executive Associate Director, Associate Director for Research Services | Review usage statistics collected from UT System libraries | Annually |
| | Steady increase in downloads at Texas Digital Library web site | Progress reports from the Texas Digital Library project | Associate Director for Digital Initiatives | Collect and review usage data for TDL | Annually |
| **Spaces** | | | | | |
| Re-engineer library public spaces to accommodate modern technologies and user behaviors. | UT Libraries maintain a customer satisfaction rating in the category of Library as Place in the LibQUAL+ survey that is equal to or higher than its peers | LibQUAL+ Survey | Assessment Coordinator | Run LibQUAL+ survey | Annually/Biennially in spring semester |
| | | | | Analyze LibQUAL+ survey results (qualitative and quantitative) | Spring and summer of survey year |
| | | | | Report on LibQUAL+ results | Fall and winter of survey year |

University of Texas Libraries Assessment Plan
7/19/06

13

http://www.lib.utexas.edu/vprovost/assessment/pdf/UTlib-Assessment-Plan_07182006.pdf

| Objective | Measure(s) of success | Assessment Method | Responsible Party | Task | Timetable |
|---|---|---|---|---|---|
| | Able to respond to 50% of customer comments received annually | Quality Council | Assessment Coordinator | Analyze comments and propose responses | Monthly |
| | | | | Report on progress at Administrative Council meetings | Biweekly |
| | Annually complete at least 10% of projects identified in the Facilities Analysis | Facilities Analysis | Associate Director for Administrative Services | Report on progress at Administrative Council meetings | Biweekly |
| Improve user seating and group study spaces within library facilities. | UT Libraries maintain a customer satisfaction rating in the category of Library as Place in the LibQUAL+ survey that is equal to or higher than its peers | LibQUAL+ Survey | Assessment Coordinator | Run LibQUAL+ survey | Annually/Biennially in spring semester |
| | | | | Analyze LibQUAL+ survey results (qualitative and quantitative) | Spring and summer of survey year |
| | | | | Report on LibQUAL+ results | Fall and winter of survey year |
| | Annually complete 5% of projects related to furniture in the Facilities Analysis | Facilities Analysis | Associate Director for Administrative Services | Report on progress at Administrative Council meetings | Biweekly |
| | Annually complete 5% of projects related to study spaces in the Facilities Analysis | Facilities Analysis | Associate Director for Administrative Services | Report on progress at Administrative Council meetings | Biweekly |

University of Texas Libraries Assessment Plan
7/19/06

14

http://www.lib.utexas.edu/vprovost/assessment/pdf/UTlib-Assessment-Plan_07182006.pdf

| Objective | Measure(s) of success | Assessment Method | Responsible Party | Task | Timetable |
|---|---|---|---|---|---|
| Redesign public areas in the Libraries to enable patrons to use spaces and services more productively; make the spaces more attractive to users and potential donors | UT Libraries maintain a customer satisfaction rating in the category of Library as Place in the LibQUAL+ survey that is equal to or higher than its peers | LibQUAL+ Survey | Assessment Coordinator | Run LibQUAL+ survey | Annually/Biennially in spring semester |
| | | | | Analyze LibQUAL+ survey results (qualitative and quantitative) | Spring and summer of survey year |
| | | | | Report on LibQUAL+ results | Fall and winter of survey year |
| | Continued use of Libraries facilities | Collection of monthly unit statistics | Assessment Coordinator | Collect unit statistics (includes gate count, circulation, reference transactions) | Due first week of the month |
| | | | | Organize and analyze unit statistics | Biannually |
| | | | | Report on unit statistics (including posting online for Libraries staff use) | Annually |
| | | Annual University Room Inventory | Associate Director for Administrative Services | Report on progress at Administrative Council meetings | Biweekly |

# Western Libraries:
# Assessment Committee Plan
# 2007/08 – 2010/11

## Introduction

In Western Libraries' strategic plan, the section on "Measuring our Success" begins, "Operating within a culture of assessment, our strategic plan incorporates and integrates performance indicators (quantitative and qualitative) for priorities in each of the key areas to provide benchmarks that will allow us to measure and validate the outcomes and impact of our proposed actions" (see Appendix attached). In a "culture of assessment" libraries make decisions based on facts, research and analysis, and use that information to plan and improve service. At the core of a culture of assessment are the users. Library users determine quality, and it is through their eyes that we learn how well we are doing and whether the library services and resources we provide add value to the institution.[1]

Within the context of Western Libraries' broader external focus on overall accountability to the University, the Assessment Committee has focused internally in this Assessment Plan to look at how we could build on the organizational infrastructure already in place for assessment to support the Libraries' decision-making and planning. The Committee chose three key priorities including LibQUAL+™ analysis, review of collections data and collections analysis, and the engagement of leaders and staff in assessment initiatives. These three priorities involve everyone in the Libraries, and the Assessment Committee is committed to providing support wherever possible.

Over the next four years Assessment Committee members will be working with leaders and staff library-wide to:

- review existing measures;
- support assessment initiatives throughout the Libraries;
- support the review of new tools both qualitative and quantitative;
- provide in depth analysis of qualitative data such as LibQUAL+™ to leaders and staff engaged in working to improve specific areas of service;
- facilitate the sharing of results of library assessment projects; and
- provide information sessions on assessment topics designed to meet particular needs as we increase our knowledge, and build methodological and analytical skills.

[1] Hiller, S. and Self, J. *Making Library Assessment Work: Practical Approaches for Developing and Sustaining Effective Assessment* Presentation given to the University of Western Ontario Libraries, March 9, 2006.

1

**Priorities, Outcomes and Action Plan for 2007/08 – 2010/11**

**Priority:**
Analysis of LibQUAL+ ™ 2007 survey of faculty, graduate and undergraduate students with priority to be given to data pertaining to graduate students and faculty 'gaps' for the purpose of achieving service improvements for those priority areas.

Outcomes:

- Canadian libraries benchmark data that will allow for Western Libraries comparison with the collective performance of G13 libraries.
- Packaged data analysis of Western Libraries data to demonstrate notable improvements since 2004 survey and/or new service 'gaps'.
- Web and print publications, similar to those used with 2004 survey, to communicate notable improvements and comparisons with G13 institutions.
- Further statistical analysis of actual data files.
- AC identification of appropriate balance among undergraduate, graduate and faculty assessment initiatives deriving from the survey.
- Assessment Committee (AC) identification of stakeholders within each priority area. Engagement with those stakeholders in developing action plans to achieve service improvements over the next four years.
- New assessment initiatives undertaken in selected areas.
- AC/stakeholder communications that report on action and successes of new assessment initiatives, relating these to the overall Western Libraries' strategic plan and multi-year budget submissions/reports to the University Administration.

Potential Action:

- Assessment Librarian to undertake SPSS training to assist in data analysis, summer 2007.
- Identify required data analysis skills and seek collaborative initiative with appropriate graduate student and faculty supervisor, summer 2007.
- Request budget support in 2007/08 budget submission for these action initiatives, September 2006.
- AC analysis of survey results to establish priorities and balance of assessment initiatives for undergraduates, graduates and faculty, by September 2007.
- Facilitate establishment of new assessment initiatives, 2007/08 – 2010/11 and assist in developing appropriate assessment methodologies to measure the success of each of the initiatives.
- Consideration of participation with next Canadian consortium LibQUAL+ ™ survey.

**Priority:**
Review of collections data sources, both quantitative and qualitative, and collections analysis tools, with priority placed on data related to graduate education and research programs. The review should include identification of data related to the expansion of

2

internationalization and interdisciplinary programs and/or any other new or emerging area of research interest for Western.

Outcomes:

- A comprehensive set of collections policies posted as appropriate.
- A clear understanding of Scholars Portal data survey and analysis initiatives and how this may facilitate assessment initiatives for Western Libraries.
- A clear understanding of availability of data using the Innovative Web Management and Electronic Resource Management Reports.
- Purchase of tools for analysis of collections, especially 'packages,' to identify redundancy and rationalize collections. Ultimately this will facilitate the switching of funds to support new and emerging areas of research.
- Identification of new or emerging areas of research priority, in association with collections analysis. Subsequent needs assessment of researchers in those areas will be required in order to ascertain collections needs specific to the area.

Potential Action:

- Ascertain timelines and scope of Scholars Portal data analysis projects.
- In collaboration with AUL (Information Resources), Library Directors and Subject Liaison Librarians, investigate the usefulness of OCLC and Bowker Collections Analysis tools and how the results of such analysis may be used in establishing priority areas for needs assessments.
- Establish a list of web sites where information may be found about successful Tri-Council grant funds awarded to Western researchers, other externally funded research grants and internally funded grants for the purpose of establishing new and emerging areas of research.
- Facilitate needs assessments for new and emerging areas of research.

**Priority:**
Engagement of leaders and staff in assessment initiatives.

Outcomes:

- Staff understanding of what is a 'culture of assessment' as a result of Assessment Committee outreach to specific groups/committees to talk about assessment initiatives and to share information, and from presentations by other staff or external experts in assessment.
- More effective data and information gathering processes.
- Needs assessments to identify unit or Libraries-wide assessment initiatives, giving priority to initiatives associated with graduate students and research teams.

Potential Action:

- Establish an outreach schedule, spring 2007.
- Undertake a review of how and why internally collected Western Libraries data are collected and used.
- Continue to maintain web site listing Western Libraries' Assessment Projects.
- New Intranet communications with a status report on Assessment Plan initiatives.

3

Appendix
(From Western Libraries Strategic Plan 2007-2008 to 2010-11)

**Measuring our Success**

Operating within a culture of assessment, our strategic plan incorporates and integrates performance indicators (quantitative and qualitative) for priorities in each of the key areas to provide benchmarks that will allow us to measure and validate the outcomes and impact of our proposed actions.

Library performance indicators for use at the institutional level include library data (if available) from the National Survey of Student Engagement (NSSE) and the Graduate and Professional Student Survey (GPSS); and the Consortium for Student Retention Data Exchange (CSRDE).

Other performance indicators currently available to the academic library community also will be used for Western Libraries: a wide range of ARL quantitative indicators; ARL LibQUAL+™ survey data providing qualitative data; Ontario Council of University Libraries (OCUL) Scholars Portal survey data supplying both quantitative and qualitative information; and, library questions on the Western Survey of Graduating Students.

As the University places a priority on library acquisitions Western Libraries places a priority on assessment of the quality, quantity and use of our collections. ARL data have been used in the past to measure the extent/quantity of collections and the size of library acquisitions expenditures. There is now a greater need for usage and qualitative data. There is also the need for collections analysis tools that will allow for ongoing analysis of existing collections, especially electronic packages, in order to avoid redundancy and to optimize spending of acquisitions funds. An important new development for Western Libraries is the OCUL plan to implement mechanisms to monitor use of Scholars Portal services and to solicit input and feedback from end users. This Scholars Portal data, in conjunction with data available from our own Web Management and Electronic Resource Management reports, will allow for further optimization of library acquisitions expenditures.

Performance indicators that will provide general measures related to the academic quality of undergraduate education outcomes include the Ontario Council of Academic Vice Presidents (OCAV) *University Undergraduate Degree Level Expectations,* in conjunction with the Association of College and Research Libraries (ACRL) Information Literacy standards. For graduate education, indicators are derived from Ontario Council on Graduate Studies (OCGS) program review and graduate degree expectations; cyclical external peer review of Departments, Schools, and Faculties; and, accreditation of professional programs, as all require and include assessment of the extent and depth of information resources in support of the programs.

Western Libraries will also use selective needs assessments targeted to specific groups using a variety of techniques: focus groups, usability studies, web surveys or one-on-one consultation. Action plans will be developed based on needs assessments followed by

4

implementation of programs with the process coming full circle by assessing how successful we are in meeting the identified needs while simultaneously identifying further improvements. Priority will be given to needs assessments and development of assessment initiatives related to graduate students and research teams, or initiatives related to identified specialist information needs (language resources, new research publications/databases, archival services, etc.).

The identification of data and testimonials related to service quality for records centre activities, alumni services, and services to the community-at-large will be selective.

Outcomes:

- A Western Libraries Assessment Plan that is an integral part of the strategic planning process.
- Performance indicators that meet both the University-level needs and our own internal needs to measure the extent and quality of collections, services, and use of space.
- Priority given to engaging our librarians/archivists and leaders in the assessment action plan in order to obtain qualitative and quantitative data so that we can make informed decisions for achieving our strategic priorities and measure how successful we are at meeting identified needs in priority areas.

# SELECTED RESOURCES

## DOCUMENTS

### Books


American Library Association. *The Reference Assessment Manual.* Ann Arbor, Michigan: Pierian Press, 1995.

Association of Research Libraries. *Proceedings of the Library Assessment Conference: Building Effective, Sustainable, Practical Assessment, September 25–27, 2007, Charlottesville, VA.* Washington, DC: Association of Research Libraries, 2007.

Booth, Andrew, and Anne Brice. *Evidence-based Practice for Information Professionals: a Handbook.* London: Facet, 2004.

Brophy, Peter. *Measuring Library Performance: Principles and Techniques*. London: Facet, 2006.

Covey, Denise T. *Usage and Usability Assessment: Library Practices and Concerns*. Washington, DC: Digital Library Federation. Council on Library and Information Resources, 2002.

De Rosa, Cathy, et al. *Perceptions of Libraries and Information Resources: A Report to the OCLC Membership.* Dublin, OH: OCLC Online Computer Library Center, 2005.

Durrance, Joan C., Karen E. Fisher, and Martin B. Hinton. *How Libraries and Librarians Help: A Guide to Identifying User-centered Outcomes*. Chicago: American Library Association, 2005.

Foster, Nancy F., and Susan Gibbons, eds. *Studying Students: The Undergraduate Research Project at the University of Rochester.* Chicago, IL: Association of College and Research Libraries, 2007. http://www.ala.org/ala/acrl/acrlpubs/downloadables/Foster-Gibbons_cmpd.pdf

Hall, Blaine H. *Collection Assessment Manual for College and University Libraries*. Phoenix, AZ: Oryx Press, 1985.

Hernon, Peter, and Ellen Altman. *Service Quality in Academic Libraries: Information Management, Policy, and Services*. Norwood, NJ: Ablex, 1996.

Hernon, Peter, and Charles McClure. *Evaluation and Library Decision Making*. Norwood, NJ: Ablex Publishing, 1990.

Hernon, Peter, and Robert E. Dugan. *An Action Plan for Outcomes Assessment in Your Library*. Chicago: American Library Association, 2002.

Kantor, Paul B. *Objective Performance Measures for Academic and Research Libraries*. Washington, DC: Association of Research Libraries, 1984.

McDonald, Joseph A., and Lynda B. Micikas. *Academic Libraries: The Dimensions of Their Effectiveness.* New Directions in Information Management, No. 32. Westport, CT: Greenwood Press, 1994.

Matthews, Joseph R. *Library Assessment in Higher Education*. Westport, CT: Libraries Unlimited, 2007.

Morgan, Steve. *Performance Assessment in Academic Libraries*. London: Mansell Publishing, 1995.

Poll, Roswitha, and Peter te Boekhorst. *Measuring Quality; Performance Measurement in Libraries.* Munich: K.G. Saur, 2007.

Stein, Joan, Martha Kyrillidou, and Denise Davis, eds. *Meaningful Measure for Emerging Realities: Proceedings of the Fourth Northumbria International Conference on Performance Measurement in Libraries and Information Services, August 12–16, 2002, Pittsburgh, PA.* Washington, DC: Association of Research Libraries, 2002.

Van House, Nancy A., Beth T. Weil, and Charles R. McClure. *Measuring Academic Library Performance: A Practical Approach.* Chicago: American Library Association, 1990.

Wallace, Danny P., and Connie J. Van Fleet. *Library Evaluation: A Casebook and Can-do Guide*. Englewood, CO: Libraries Unlimited, 2001.

Zweizig, Douglas L. *The Tell It! Manual: The Complete Program for Evaluating Library Performance*. Chicago: American Library Association, 1996.

## Journal Articles

Alemna, Anaba A. "Library Evaluation: Some Options for Academic Libraries." *Aslib Proceedings* 51, no. 7 (July/August 1999): 243–46.

Baker, Ronald L. "Evaluating Quality and Effectiveness: Regional Accreditation Principles and Practices." *Journal of Academic Librarianship* 28, no. 1/2 ( January/March 2002): 3–7.

Bauer, Kathleen. "Resources for Library Assessment." *College and Research Libraries News* 62, no. 1 (January 2001): 12–14.

Blixrud, Julia C. "Mainstreaming New Measures." *ARL: A Bimonthly Report on Research Library Issues and Actions from ARL, CNI, and SPARC*, no. 230/231 (October/December 2003): 1–8.

Edgar, William B. "Questioning LibQUAL+™: Expanding Its Assessment of Academic Library Effectiveness." *portal: Libraries and the Academy* 6, no. 4 (October 2006): 455–65.

Fraser, Bruce T., Emily H. Leahy, and Charles R. McClure. "Toward a Framework for Assessing Library and Institutional Outcomes." *portal: Libraries and the Academy* 2, no. 4 (October 2002): 505–28.

Goodall, Deborah. L. "Performance Measurement: a Historical Perspective." *Journal of Librarianship* 20, no. 2 (April 1988): 128–44.

Hernon, Peter, and Danuta Nitecki. "Service Quality: A Concept Not Fully Explored." *Library Trends* 49, no. 4 (Spring 2001): 687–708.

Hiller, Steve, and Jim Self. "From Measurement to Management: Using Data Wisely for Planning and Decision-Making." *Library Trends* 53, no. 1 (Summer 2004): 129–55.

Kyrillidou, Martha. "From Input and Output Measures to Quality and Outcome Measures, or, From the User in the Life of the Library to the Library in the Life of the User." *Journal of Academic Librarianship* 28, no. 1/2 (January 2002): 42–46.

Laeven, Hubert, and Anja Smit. "A Project to Benchmark University Libraries in the Netherlands." *Library Management* 24, no. 6/7 (2003): 291–304.

Lakos, Amos, and Shelley Phipps. "Creating a Culture of Assessment: A Catalyst for Organizational Change." *portal: Libraries and the Academy* 4, no. 3 (July 2004): 345–61.

Lindauer, Bonnie G. "Defining and Measuring the Library's Impact on Campus Wide Outcomes." *College & Research Libraries* 59, no. 6 (November 1998): 546–70.

McClure, Charles R. "A View from the Trenches: Costing and Performance Measures for Academic Library Public Services." *College & Research Libraries* 47 (July 1986): 323–36.

Nicholson, Scott. "A Conceptual Framework for the Holistic Measurement and Cumulative Evaluation of Library Services." *Journal of Documentation* 60, no. 2 (2004): 164–82.

Nicholson, Scott. "Proof in the Pattern: Librarians Follow the Corporate Sector Toward More Data-driven Management. (Managing by Numbers)." *Library Journal* 131, no. 1 (Winter 2006): S2–S6.

Nitecki, Danuta A. "Quality Assessment Measures in Libraries." *Advances in Librarianship* 25. New York, NY: Academic Press, 2001: 133–62.

Powell, Ronald R. "Evaluation Research: An Overview." *Library Trends* 55, no. 1 (Summer 2006): 102–20.

Pritchard, Sarah M. "Determining Quality in Academic Libraries." *Library Trends* 44, no. 3 (Winter 1996): 572–94.

Rowley, Jennifer. "Making Sense of the Quality Maze: Perspectives for Public and Academic Libraries." *Library Management* 26, no. 8/9 (2005): 508–18.

Saunders, Laura. "Regional Accreditation Organizations' Treatment of Information Literacy: Definitions, Collaboration, and Assessment." *Journal of Academic Librarianship* 33, no. 3 (May 2007): 317–26.

St. Clair, Gloriana. "Improving Quality: Organizational Benefits of Total Quality Management." *College & Research Libraries* 54 (September 1993): 370–72.

Shi, Xi, and Sarah Levy. "A Theory-guided Approach to Library Services Assessment." *College & Research Libraries* 66, no. 3 (May 2005): 266–77.

Stubbs, Kendon. "University Libraries: Standards and Statistics." *College & Research Libraries* 42 (November 1981): 528–38.

Weiner, Sharon A. "Library Quality and Impact: Is There a Relationship Between New Measures and Traditional Measures?" *Journal of Academic Librarianship* 31, no. 5 (September 2005): 432–37.

Winkworth, Ian. "Quality Issues and Performance Indicators in Academic Libraries (1)." British Library. Research and Innovation Report 92, 1998: 1–6.

## Library Assessment Web Sites

University of Alberta
http://www.library.ualberta.ca/survey_result/index.cfm#desir

Boston College Libraries
http://www.bc.edu/libraries/about/assessment/

University of California-Davis Libraries
http://www.lib.ucdavis.edu/ul/about/aur/

University of Connecticut Libraries
http://www.lib.uconn.edu/about/administration/surveys/

Cornell University Library
http://www.library.cornell.edu/iris/research/assessment/

University of Georgia Libraries
http://dataserv.libs.uga.edu/assessment/index.html

Indiana University-Perdue University-Indianapolis Library
http://www.ulib.iupui.edu/prod/portfolio/plan

University of Maryland Libraries
http://www.lib.umd.edu/PASD/MIS/larc/

University of Massachusetts Amherst Libraries
http://www.library.umass.edu/assessment/index.html

Rutgers University Libraries
http://www.libraries.rutgers.edu/rul/staff/planning/planning.shtml

University of Texas-Austin Libraries
http://www.lib.utexas.edu/vprovost/assessment/index.html

University of Waterloo Library
http://www.lib.uwaterloo.ca/staff/community/index.html

Yale University Library
http://www.library.yale.edu/assessment/

## Other Resources

American Library Association: Office for Research and Statistics
http://www.ala.org/ala/ors/statsaboutlib/statisticsabout.htm

Association of College and Research Libraries. Academic Library Statistics
http://www.ala.org/ala/acrl/acrlpubs/acadlibrarystats/academiclibrary.cfm

Association of Research Libraries
http://www.arl.org/arl/

Evidence-Based Library Practice
http://conferences.alia.org.au/ebl2005/
http://www.eblip4.unc.edu/

Library Assessment Conference
http://www.libraryassessment.org/

National Center for Education Statistics. Library Statistics Program
http://nces.ed.gov/surveys/libraries/

Northumbria International Conference on Performance Measurement in Libraries and Information Services http://www.lib.sun.ac.za/Northumbria7/

SCONUL (Society of College, National and University Libraries)
http://vamp.diglib.shrivenham.cranfield.ac.uk/

Special Libraries Association. Library Assessment and Benchmarking Institute
http://www.sla.org/content/learn/onlinepres/LAB2002/index.cfm

State Library Statistics Sources
http://web.syr.edu/~jryan/infopro/stats1.html

Note: All URLs accessed November 14, 2007.

# SPEC KIT TITLE LIST

SP303 Library Assessment
SP302 Managing Public Computing
SP301 Liaison Services
SP300 Open Access Resources
SP299 Scholarly Comm. Educ. Initiatives
SP298 Metadata
SP297 Library Development
SP296 Public Services in Special Collections
SP295 Remote Shelving Facilities
SP294 Managing Digitization Activities
SP293 External Review for Promo & Tenure
SP292 Institutional Repositories
SP291 Spatial Data Collections & Services
SP290 Access Services
SP289 Managing Large Projects
SP288 Scanning Services for Library Users
SP287 Instructional Improvement Programs
SP286 Collab for Dist Learn Info Lit Instr
SP285 Lib Svcs in Non-Library Spaces
SP284 Security in Special Collections
SP283 Grant Coordination
SP282 Managing Electronic Resources
SP281 The Information Commons
SP280 Library User Surveys
SP279 Evaluating Library Instruction
SP278 Library Patron Privacy
SP277 Lib Pub Acc Workstation Auth
SP276 Recruitment and Retention
SP275 Laptop Computer Services
SP274 Data Mining & Warehousing
SP273 Chat Reference
SP272 Insuring & Valuing Res Lib Coll
SP271 Lib Systems Office Organization
SP270 Core Competencies
SP269 Integrating Preserv Activities
SP268 Reference Statistics
SP267 User Authentication
SP266 Staffing the Library Website
SP265 Instructional Support Services
SP264 Extended Library Hours
SP263 Numeric Data Services
SP262 Preservation & Digitization
SP261 Post-Tenure Review
SP260 Interview Process
SP259 Fee-based Services
SP258 Corporate Annual Reports
SP257 MLS Hiring Requirement
SP256 Changing Roles of Lib Profs
SP255 Branch Libs/Discrete Collectns
SP254 Managing Printing Services
SP253 Networked Info Services
SP252 Supprt Staff Classifictn Studies
SP251 Electronic Reference Service
SP250 TL10: Educating Faculty
SP249 Catalogng of Resrces Digitized
SP248 Licensing of Electronic Prodcts
SP247 Management of Lib Security
SP246 Web Page Devel & Managmnt
SP245 Electronic Reserves Operations
SP244 TL 9: Renovation & Reconfigur
SP243 TL 8: Users with Disabilities
SP242 Library Storage Facilities
SP241 Gifts and Exchange Function
SP240 Marketing and PR Activities
SP239 Mentoring Programs in ARL
SP238 ARL GIS Literacy Project
SP237 Managing Food and Drink
SP236 TL 7: E-Theses/Dissertations
SP235 Collaborative Coll Management
SP234 TL 6: Distance Learning
SP233 ARL in Extension/Outreach
SP232 Use of Teams in ARL
SP231 Cust Service Programs in ARL
SP230 Affirmative Action in ARL
SP229 Evaluating Acad Libr Dirs
SP228 TL 5: Preserving Digital Info
SP227 Org of Doc Coll & Svcs
SP226 TL 4: After the User Survey
SP225 Partnerships Program
SP224 Staff Training & Development
SP223 TL 3: Electronic Scholarly Pubn
SP222 Electronic Resource Sharing
SP221 Evol & Status of Approval Plans
SP220 Internet Training
SP219 TL 2: Geographic Info Systems
SP218 Info Technology Policies
SP217 TL 1: Electronic Reserves
SP216 Role of Libs in Distance Ed
SP215 Reorg & Restructuring
SP214 Digit Tech for Preservation
SP213 Tech Svcs Workstations
SP212 Non-Librarian Professionals
SP211 Library Systems Office Org
SP210 Strategic Planning
SP209 Library Photocopy Operations
SP208 Effective Library Signage
SP207 Org of Collection Develop
SP206 Faculty Organizations
SP205 User Surveys in ARL Libs
SP204 Uses of Doc Delivery Svcs
SP203 Reference Svc Policies
SP202 E-journals/Issues & Trends
SP201 E-journals/Pol & Proced
SP200 2001: A Space Reality
SP199 Video Collect & Multimedia
SP198 Automating Preserv Mgt
SP197 Benefits/Professional Staff
SP196 Quality Improve Programs
SP195 Co-op Strategies in Foreign Acqs
SP194 Librarian Job Descriptions
SP193 Lib Develop & Fundraising
SP192 Unpub Matls/Libs, Fair Use
SP191 Prov Pub Svcs Remote User
SP190 Chang Role of Book Repair
SP189 Liaison Svcs in ARL Libs
SP188 Intern, Residency & Fellow
SP187 ILL Trends/Staff & Organ
SP186 Virtual Library
SP185 System Migration
SP184 ILL Trends/Access
SP183 Provision of Comp Print Cap
SP182 Academic Status for Libns
SP181 Perf Appr of Collect Dev Libn
SP180 Flexible Work Arrangemts
SP179 Access Services Org & Mgt
SP178 Insuring Lib Colls & Bldgs
SP177 Salary Setting Policies
SP176 Svcs for Persons w/Disabilities
SP175 Scholarly Info Centrs
SP174 Expert Systems
SP173 Staff Recognition Awards
SP172 Information Desks
SP171 Training of Tech Svc Staff
SP170 Organization Charts
SP169 Mgt of CD-ROM
SP168 Student Employment
SP167 Minority Recruitment
SP166 Materials Budgets
SP165 Cultural Diversity
SP164 Remote Storage
SP163 Affirmative Action
SP162 Audiovisual Policies
SP161 Travel Policies
SP160 Preservation Org & Staff
SP159 Admin of Lib Computer Files
SP158 Strategic Plans
SP157 Fee-based Services
SP156 Automating Authority Control
SP155 Visiting Scholars/Access
SP154 Online Biblio Search
SP153 Use of Mgt Statistics
SP152 Brittle Books Program
SP151 Qualitative Collect Analysis
SP150 Bldg Security & Personal Safety
SP149 Electronic Mail
SP148 User Surveys
SP147 Serials Control/Deselection
SP146 Lib Dev Fund Raising Capabilit
SP145 Lib Publications Programs
SP144 Building Use Policies
SP143 Search Proced Sr LibAdmin
SP142 Remote Access Online Cats
SP141 Approval Plans
SP140 Performance Appraisal
SP139 Performance Eval: Ref Svcs
SP138 University Copyright
SP137 Preservation Guidelines
SP136 Managing Copy Cataloging
SP135 Job Analysis
SP134 Planning Mgt Statistics
SP133 Opt Disks: Storage & Access
SP132 Library-Scholar Communication
SP131 Coll Dev Organization
SP130 Retrospective Conversion
SP129 Organization Charts

SP128 Systems File Organization
SP127 Interlibrary Loan
SP126 Automated Lib Systems
SP125 Tech Svcs Cost Studies
SP124 Barcoding of Collections
SP123 Microcomp Software Policies
SP122 End-User Search Svcs
SP121 Bibliographic Instruction
SP120 Exhibits
SP119 Catalog Maintenance Online
SP118 Unionization
SP117 Gifts & Exchange Function
SP116 Organizing for Preservation
SP115 Photocopy Services
SP114 Binding Operations
SP113 Preservation Education
SP112 Reorg of Tech and Pub Svcs
SP111 Cooperative Collection Dev
SP110 Local Cataloging Policies
SP109 Staff Training for Automation
SP108 Strategic Planning
SP107 University Archives
SP106 Electronic Mail
SP105 Nonbibliographic Dbases
SP104 Microcomputers
SP103 Asst/Assoc Dir Position
SP102 Copyright Policies
SP101 User Studies
SP100 Collection Security
SP099 Branch Libraries
SP098 Telecommunications
SP097 Building Renovation
SP096 Online Catalogs
SP095 Lib Materials Cost Studies
SP094 Fund Raising
SP093 User Instructions for Online Cats
SP092 Interlibrary Loan
SP091 Student Assistants
SP090 Integrated Lib Info Systems
SP089 Tech Svcs Cost Studies
SP088 Corporate Use of Research Libs
SP087 Collect Descript/Assessment
SP086 Professional Development
SP085 Personnel Classification Sys
SP084 Public Svcs Goals & Objectvs
SP083 Approval Plans
SP082 Document Delivery Systems
SP081 Services to the Disabled
SP080 Specialty Positions
SP079 Internships/Job Exchanges
SP078 Recruitment-Selection
SP077 Use of Small Computers
SP076 Online Biblio Search Svcs
SP075 Staff Development
SP074 Fees for Services
SP073 External User Services
SP072 Executive Review
SP071 User Surveys: Eval of Lib Svcs
SP070 Preservation Procedures
SP069 Prep Emergencies/Disasters
SP068 AACR2 Implement Studies
SP067 Affirm Action Programs
SP066 Planning Preserv of Lib Mat
SP065 Retrospective Conversion
SP064 Indirect Cost Rates
SP063 Collective Bargaining
SP062 Online Biblio Search Svcs
SP061 Status of Librarians
SP060 Lib Materials Cost Studies
SP059 Microform Collections
SP058 Goals & Objectives
SP057 Special Collections
SP056 External Communication
SP055 Internl Com/Staff & Super Role
SP054 Internal Com/Policies & Proced
SP053 Performance Appraisal
SP052 Cost Studies & Fiscal Plan
SP051 Professional Development
SP050 Fringe Benefits
SP049 Use of Annual Reports
SP048 External Fund Raising
SP047 Automated Cataloging
SP046 Plan Future of Card Catalog
SP045 Changing Role Personnel Officer
SP044 Automated Acquisitions
SP043 Automated Circulation Sys
SP042 Resource Sharing
SP041 Collection Assessment
SP040 Skills Training
SP039 Remote Storage
SP038 Collection Dev Policies
SP037 Theft Detection & Prevent
SP036 Allocation Materials Funds
SP035 Preservation of Lib Materials
SP034 Determin Indirect Cost Rate
SP033 Intergrat Nonprint Media
SP032 Prep, Present Lib Budget
SP031 Allocation of Resources
SP030 Support Staff, Student Assts
SP029 Systems Function
SP028 Gifts & Exchange Function
SP027 Physical Access
SP026 Bibliographic Access
SP025 User Statistics and Studies
SP024 User Surveys
SP023 Grievance Policies
SP022 Private Foundations
SP021 Paraprofessionals
SP020 Managerial Technical Specialists
SP019 Staff Allocations
SP018 Staff Development
SP017 Library Instruction
SP016 Reclassification
SP015 Goals & Objectives
SP014 Performance Review
SP013 Planning Systems
SP012 Acquisition Policies
SP011 Collection Development
SP010 Leave Policies
SP009 Tenure Policies
SP008 Collective Bargaining
SP007 Personnel Class Schemes
SP006 Friends of the Lib Organizations
SP005 Performance Review
SP004 Affirmative Action
SP003 A Personnel Organization
SP003 Status of Librarians
SP002 Personnel Survey (flyer only)
SP001 Organization Charts

**SPEC Kit Price Information**

Individual Kits: $35 ARL members/$45 nonmembers, plus shipping and handling.
Individual issues of the Transforming Libraries (TL) subseries: $28, plus shipping and handling.

**Shipping & Handling**

U.S.: UPS Ground delivery, $10 per publication.
Canada: UPS Ground delivery, $15 per publication
International and rush orders: Call or e-mail for quote.

**Payment Information**

Make check or money order payable in U.S. funds to the **Association of Research Libraries**, Federal ID #52-0784198-N.
MasterCard and Visa accepted.

**Send orders to:** ARL Publications Distribution Center, P.O. Box 531, Annapolis Junction, MD 20701-0531
phone (301) 362-8196; fax (301) 206-9789; e-mail pubs@arl.org

**Order online at:** http://www.arl.org/resources/pubs/index.shtml

Made in the USA
Coppell, TX
10 November 2026

11835556R00107